74th Annual Meeting of the Labor and Employment Relations Association and LERA@ASSA Meeting 2022

Labor and Employment Relations Association Series

Online
7 – 9 January 2022
2 – 5 June 2022

ISBN: 978-1-7138-7083-8

Printed from e-media with permission by:

Curran Associates, Inc.
57 Morehouse Lane
Red Hook, NY 12571

Some format issues inherent in the e-media version may also appear in this print version.

Printed with permission by Curran Associates, Inc. (2023)

Labor and Employment Relations Association (LERA)
University of Illinois at Urbana-Champaign
121 Labor and Employment Relations Bldg.,
504 E. Armory Ave.
Champaign IL 61820
USA

Phone: (217) 333-0072
Fax: (217) 265-5130

LERAoffice@illinois.edu

Additional copies of this publication are available from:

Curran Associates, Inc.
57 Morehouse Lane
Red Hook, NY 12571 USA
Phone: 845-758-0400
Fax: 845-758-2633
Email: curran@proceedings.com
Web: www.proceedings.com

LABOR AND EMPLOYMENT RELATIONS ASSOCIATION SERIES

Proceedings of the LERA 2022 Meetings

LERA@ASSA Meeting
January 7–9, 2022
Virtual

LERA 74th Annual Meeting
June 2–5, 2022
Virtual

OFFICERS OF LERA FOR 2021–2022

PRESIDENT
Wilma Liebman, Former Chair NLRB

PRESIDENT ELECT
Paul Clark, Pennsylvania State University

PAST PRESIDENT
Adrienne Eaton, Rutgers University

SECRETARY-TREASURER
Andrew Weaver, University of Illinois at Urbana-Champaign

EDITOR-IN-CHIEF
J. Ryan Lamare, University of Illinois at Urbana-Champaign

CHAPTER ADVISORY CHAIR:
William Canak, Middle Tennessee State University

EXECUTIVE DIRECTOR
Emily Smith, University of Illinois at Urbana-Champaign

REGIONAL VICE PRESIDENTS
Jim Pruitt, Kaiser Permanente (West Region)
Beverly E. Harrison, Arbitrator/Mediator (East Region)
Robert Chiaravalli, Strategic Labor and HR LLC (Mid Region)

EXECUTIVE BOARD MEMBERS:
Daniel Altchek, Saul Ewing Arnstein & Lehr LLP
Peter Berg, Michigan State University
Julie Farb, AFL-CIO
Janet Gillman, Oregon Employment Relations Board
Shannon Gleeson, Cornell University
Kati Griffith, Cornell University
Quinton Herbert, City of Baltimore
Erin Johansson, Jobs With Justice
Tamara Lee, Rutgers University
Kevin Legel, Ford Motor Co.
Deborah Mueller, CSEA Local 1000 AFSCME
Javier Ramirez, Federal Mediation and Conciliation Service
Christine Riordan, University of Illinois at Urbana-Champaign
Jake Rosenfeld, Washington University in St. Louis
Palak Shah, National Domestic Workers Alliance
Marc Weinstein, Florida International University

LEGAL COUNSEL:
Steven B. Rynecki, von Briesen, Purtell & Roper, S.C.

PAST PRESIDENTS OF LERA:

1948—Edwin E. Witte
1949—Sumner H. Slichter
1950—George W. Taylor
1951—William Leiserson
1952—J. Douglas Brown
1953—Ewan Clague
1954—Clark Kerr
1955—Lloyd G. Reynolds
1956—Richard A. Lester
1957—Dale Yoder
1958—E. Wight Bakke
1959—William Haber
1960—John T. Dunlop
1961—Philip Taft
1962—Charles A. Myers
1963—William F. Whyte
1964—Solomon Barkin
1965—Edwin Young
1966—Arthur M. Ross
1967—Neil Chamberlain
1968—George P. Shultz
1969—Frederick Harbison
1970—Douglass V. Brown
1971—George Hildebrand
1972—Benjamin Aaron
1973—Douglas H. Soutar
1974—Nathaniel Goldfinger
1975—Gerald G. Somers
1976—Irving Bernstein
1977—F. Ray Marshall
1978—Charles C. Killingsworth
1979—Jerome M. Rosow
1980—Jack Barbash
1981—Rudolph A. Oswald
1982—Milton Derber
1983—Jack Stieber
1984—Wayne L. Horvitz
1985—Everett Kassalow
1986—Lloyd Ulman
1987—Michael H. Moskow
1988—Phyllis A. Wallace
1989—Joyce D. Miller
1990—Robert B. McKersie
1991—James L. Stern
1992—Ernest J. Savoie
1993—George Strauss
1994—Lynn R. Williams
1995—Walter Gershenfeld
1996—Hoyt N. Wheeler
1997—Francine D. Blau
1998—F. Donal O'Brien
1999—Thomas A. Kochan
2000—Sheldon Friedman
2001—Magdalena Jacobsen
2002—John F. Burton, Jr.
2003—Paula B. Voos
2004—Marlene Heyser
2005—Stephen R. Sleigh
2006—David B. Lipsky
2007—Eileen B. Hoffman
2008—Anthony Oliver, Jr.
2009—Joel Cutcher-Gershenfeld
2010—Eileen Appelbaum
2011—Gordon Pavy
2012/13—David Lewin
2013/14—Lisa Lynch
2014/15—Martin Mulloy
2015/16—Bonnie Castrey
2016/17—Janice Bellace
2017/18—Harry Katz
2018/19—Kris Rondeau
2019/20—Dennis Dabney
2020/21—Adrienne Eaton

LABOR AND EMPLOYMENT RELATIONS

ASSOCIATION SERIES

Proceedings of the LERA 2022 Meetings

LERA@ASSA Meeting
January 7–9, 2022, Virtual
(in conjunction with ASSA/AEA)

and

LERA 74th Annual Meeting,
June 2–5, 2022, Virtual

J. Ryan Lamare, Editor-in-Chief

First Edition

LABOR AND EMPLOYMENT RELATIONS ASSOCIATION SERIES
(Formerly the Industrial Relations Research Association)
Proceedings of the Annual Meeting
Annual Research Volume
Perspectives on Work Magazine

Inquiries and other communications regarding membership, meetings, publications, and general affairs of the Association, as well as notice of address changes, should be addressed to the LERA office.

LABOR AND EMPLOYMENT RELATIONS ASSOCIATION
University of Illinois at Urbana-Champaign School of Labor and Employment Relations
121 Labor and Employment Relations Building, MC-504
504 East Armory Avenue, Champaign, IL 61820
Telephone: 217/333-0072 Fax: 217/265-5130 E-mail: LERAoffice@illinois.edu
Website: http://lera.memberclicks.net

CONTENTS

LERA 74th ANNUAL MEETING, JUNE 2–5, 2022, VIRTUAL

VI.
LERA BEST POSTERS
Hye Jin Rho, Presiding

VII.
2022 LERA LIFETIME ACHIEVEMENT AWARD

LERA@ASSA MEETING, JANUARY 7–9, 2022, VIRTUAL

VIII.
LERA BEST PAPERS I: GENDER AND PAY
Claudia Olivetti, Presiding

IX.
LERA BEST PAPERS II: JOB QUALITY
John Schmitt, Presiding

X.
LERA/IAFFE JOINT SESSION: DIVERSITY IN THE LABOR MARKET
Takao Kato, Presiding

XI.
LERA ANNUAL REPORTS

I. Presidential Interview

The State of Unionism Today

WILMA LIEBMAN
Former Chair, National Labor Relations Board

Editor's note: The following is the text of a prerecorded conversation between Wilma Liebman, LERA president and past National Labor Relations Board chair, and Josh Eidelson, Bloomberg labor reporter, televised on June 4 at the LERA 2022 virtual annual meeting.

LIEBMAN: I am delighted that you agreed to join me in this conversation. We've been talking for more than a decade. You were a union organizer at Yale College, and your career as a journalist covering labor and work has blossomed. You report so insightfully—and have learned so much directly from workers, especially young workers—about what's happening all over the nation in low-wage to high-tech workplaces and about the range of dynamic activism. You've been following labor policy, especially at the National Labor Relations Board, during different administrations and in some statehouses around the country.

I now observe all of this from a distance, no longer as a policy maker but with the perspective of a long career in labor law/relations, first as an advocate and then in government. I see what's going on today with both dread and fascination—dread about the divides, the hate, the violence, the pain and despair, about democratic norms eroding and democratic institutions under great stress, but also with fascination about this historic moment of dynamism in both thought and action not seen perhaps in a century. I thought that a conversation from our different—and across generations—vantage points might be of interest to our audience and fun, at least for us.

EIDELSON: As you mentioned, your career stretches throughout government and outside of government and across a slew of organizations. You have found a way since leaving the National Labor Relations Board to be involved in some of the most interesting things happening in the workplace. I want to start with your experience working on models that are alternatives, such as the effort to have a works council with a union at Volkswagen and more recently your work on employee ownership. Have those shifted your perspectives from how you saw things when you were on the NLRB; do you see them as alternatives to NLRA unionism, or as something else?

LIEBMAN: I have been very lucky since leaving the NLRB. It is true that, up until then, my focus had been on traditional labor law, first as a union lawyer and then during my nearly 14 years as a member and then chair of the NLRB. My focus since then has been on some really interesting projects that push the boundaries of traditional labor law or things that are occurring *notwithstanding* labor law and its limits. You mentioned two things.

About eight or nine years ago, I was asked by the United Auto Workers to work with them when they were engaged in an organizing campaign at the Volkswagen plant in Chattanooga, Tennessee. They were interested, and in dialogue with Volkswagen, about trying to set up a German-style works council, if they were successful in organizing the plant.

They consulted me because they wanted to know if there was a chance that Volkswagen could set up a works council on its own, without union representation, consistent with U.S. labor law. My conclusion was that under existing labor law they could not do that. Unfortunately, though, the organizing campaign did not succeed. There was a lot of outside interference from various politicians in the state.

There was a second election that the union also lost, more narrowly than the first. So far, this experiment hasn't materialized, but, at the time, the attempt by the United Auto Workers, in dialogue with Volkswagen, sparked a lot of interest and probably a renewed interest in the German model of co-determination, with its enterprise-level works councils and worker representation on supervisory boards, along with, in the dual model, the unions themselves negotiating the economic pieces of the relationship, with companies and across sectors. There has been a lot of interest in this model.

More recently, I've been involved in some efforts at worker ownership. I first became knowledgeable about these initiatives when I spent some time at Rutgers and learned from their experts on worker ownership. Lately I've been asked to sit on a board of a brand-new nonprofit called Ownership Works, which is trying to promote worker equity shares through private-equity deals.

What's most interesting to me about both of these efforts is that they exemplify the recent reimagining of "industrial democracy." A few years ago, I was asked to contribute to a conference held in Paris on the occasion of the 100th anniversary of the International Labour Organisation. I wrote a paper on the "labor question" that people were debating a century ago; the answer to the labor question was generally said to be industrial democracy.

But what industrial democracy meant was subject to wide interpretation among different segments of labor and social movements—some more radical than others—and among employers themselves, who talked about industrial democracy but had their own ideas about what that meant. They were hoping, of course, to have some kind of industrial democracy short of either independent labor unions or government intervention.

The range of approaches included collective bargaining, which the AFL-CIO unions promoted. Some, like the Knights of Labor and other groups, promoted worker ownership of factories. Employers, for their part, were instituting employee-representation plans, or works council–type entities.

During World War I, the government did intervene to try to maintain labor peace, and shop committees were set up in industries all over the nation. Unions thought that, even though these were non-union, there was some hope that they would turn into union-represented organizations. That wasn't to happen, however. At the end of the war, the government retrenched, and there was repression of labor by government and by employers.

During the 1920s, the welfare-capitalism model took hold, and employers once again instituted a variety of worker councils, representation plans, profit sharing, and the like. But those worker councils, some similar to German works councils, were eventually outlawed by the 1935 Wagner Act, which adopted the collective bargaining model as our nation's vision of industrial democracy, and that is what prevails in the law today.

So, these recent efforts are essentially, I think, an attempt to reimagine different, earlier forms of industrial democracy, and that's the piece that I find so exciting.

EIDELSON: Speaking of this moment we're in, you referenced, I think, this upsurge that we're seeing around the country in organizing in unexpected places at some of the most prominent, most powerful companies in the United States. It might have seemed safe a few years ago to assume that a NLRB vote at a Starbucks café or an Amazon warehouse would at best for the labor movement seem like a moral victory, a noble failure. But we've now actually seen victories at both companies.

How big a deal is this long term, and particularly how significant can the impact of these kinds of wins be in the absence of federal labor law reform? Is there still a Catch-22 between an upsurge in worker organizing and a transformation of the labor law? Or has that Catch-22 somehow been blown up by some of these victories we've been seeing recently?

LIEBMAN: That's a multipart question. Let me just start with part of your first question that I didn't answer: whether I see these other models as alternatives to traditional labor unions. I do not see them as alternatives; I see them as complementary. We have a very complex economy, and I think there's room for a lot of different forms of worker voice, worker engagement, worker power. And I think that's what is most interesting about this moment: there is a dynamism of both activism and intellectual thought that we really haven't seen, perhaps in 100 years.

Certainly, the worker activism we've seen accelerated is remarkable. It's really a historic moment. I think it began about ten years ago with Occupy Wall Street, then with the Fight for $15, which started small in New York City and now has grown all over the country and globally. This activism has certainly accelerated during the pandemic with the intersecting crises: the pandemic and attendant economic consequences, the racial justice crisis, the climate crisis, the democracy crisis. All of these intersecting and cascading crises have led to a growing awareness by many people, particularly workers, of the inadequacies of our social contract, of the weakness of our social safety net—that workers have become terribly disempowered. They've lost a voice, and tremendous discontent and frustration has bubbled up.

We've seen this not only with the so-called Great Resignation, with people leaving jobs and moving to other jobs, but also with the demands for higher wages. Some of that, of course, is a result of the Fight for $15, which, while it didn't result in widespread unionization, did spur minimum wage hikes all over the country. I think what we are seeing now is very complex.

The worker organizing we see is both with traditional unions and with start-up organizations like the Amazon win at Staten Island. But even earlier, the year before, the loss at the Amazon Bessemer, Alabama, plant, I think, was significant.

Obviously, there's nothing like winning—and winning is contagious, as we're seeing at Starbucks. But the media attention the Bessemer campaign received was striking. And I have to footnote here that I think reporters like you and others who are covering these workplace stories are really a major factor in the growing awareness on the part of the public about the needs of workers. Ten years ago, you didn't see the newspapers and the media plastered with stories about what's going on in the workplace. So, I think we're seeing all different kinds of worker organizing, some of it with the goal of collective bargaining, some of it not yet. But this is something that people ignore at their peril, especially in the corporate world.

This is all happening without labor law reform. I have always thought that labor law reform would happen only with a reinvigorated worker movement or labor movement. I think it's going to go that direction rather than vice versa. It's a Catch-22 in the sense that labor law reform would presumably also catalyze more union organizing. But I don't think, given our deep partisan divides and our deep divides on these kinds of issues, not to mention the war on government that is so powerful in our country, that we will see significant labor law reform absent sustained, powerful worker activism.

EIDELSON: I want to turn to the role of your old home, the NLRB, in all of this right now following President Biden's appointments. We have a general counsel, Jennifer Abruzzo, who's taken an unusually ambitious approach to trying to change precedent, not only from Trump, but going back to Eisenhower and presidents in between. We have a Democratic majority signaling they take a broader view of what the NLRA protects. What do you expect we will see by the time the Republicans again are back in power and able to replace these folks? What do you think can or should get done?

LIEBMAN: Well, that depends on when the Republicans come back in power, *if* they come back in power. But what we are seeing, obviously, is a very aggressive, ambitious general counsel. Some of Abruzzo's approach is a result of the profound and sweeping precedent changes during the Trump administration. Some of it, I think, is occasioned by what we just talked about: the amazing activism and dynamism going on in workplaces. And also a younger generation of workers, with all that represents—both the diversity, changing needs and demands, the desire for a voice. Yes, they want economic improvements, but they also want a voice. They want dignity in the workplace. So, I think the dynamics of this, with younger workers—and I'd be interested in in your perspectives on this because you talk to a lot of them, I know—is really making for some interesting cases coming before the NLRB.

So far, the board itself has not issued any precedent-setting cases. They have had a lot thrown at them, not to mention dealing with the rulemakings that the Trump board engaged in. Rulemaking is time consuming and labor intensive. But for sure what I think we will see coming out of this board is a much more dynamic, living view of the law than we've seen, certainly during the Trump administration.

When I talk about a living law, it's the Supreme Court Justice Louis Brandeis's notion of keeping the law alive despite changing circumstances. Of course, the board has to act within certain constraints, such as an aging statute, not significantly amended since 1947, and decades of precedent they have to navigate. And they have to be mindful of a very skeptical and even hostile judiciary, even more so as a result of the Trump appointments. I think any board committed to a dynamic interpretation of the law can make very meaningful change, but they can't make radical change without statutory change, and the board, of course, is known for flip-flopping or policy oscillation.

We'll see that oscillation again absent any kind of real statutory change. But there are a few areas of the law in which I'm sure they will make some noticeable rulings. Those would include statutory coverage issues: who's an employee; who's an employer. These are always vexing issues with legal tests that sometimes flip-flop, and that has included the employee status issue of graduate teaching assistants, for example, not to mention who's excluded from coverage as independent contractors—the Uber issue. With respect to remedies for unfair labor practices, and the labor law's famously weak remedies, I'm sure they will do their best, within the constraints of the law, to try to impose meaningful remedies.

And then perhaps some of the most interesting cases turn on the scope of concerted activity protected by the law. Workers are, I think, pushing the boundaries of what the board has traditionally considered to be within the scope of protected, concerted activity. I think we'll see some interesting decisions. You've been covering one interesting trial that maybe we could talk about [the Whole Foods case involving whether wearing Black Lives Matter facial masks is protected activity]. I think also that the board certainly will make some rulings or issue rulemakings with respect to NLRB election procedures and the oversight of campaign interference.

EIDELSON: Reading one of Nelson Lichtenstein's books recently, I was surprised to learn just how common industrial democracy was in U.S. mainstream political discourse a century ago, in a way it isn't now. And it's been interesting for me traveling around the country talking to Starbucks workers to see what a theme of the campaign with regard to democracy and having voice in the workplace has been.

Many of these workers have been insistent that there is no particular issue, or even set of issues, that the company could resolve that would sufficiently mitigate their sense that they need a seat at the table. And they point frequently to Starbucks' invocation of an empty chair that's at the table representing them and how they're in collective bargaining, trying to negotiate an actual seat in that actual chair with the company. From your perspective after years since leaving the board, are there particular issues that you wish you had been able to tackle more while you were there or to approach in a different way?

LIEBMAN: The coverage issues always interested me because I was quite struck during my time at the board about how ingenious the business world was in the way they structured their operations, and particularly the way they arranged for work to be done. That often led to issues of whether they were the employer of the workers in question, whether they had to bargain with them, whether they were joint employers, whether they were considered an alter ego or a single employer, in the parlance of the law, or whether the workers were independent contractors or were their employees.

I would have liked to have been able to decide more of those questions clarifying the controlling legal tests. I did write dissenting opinions on some of those issues. After I left the board, some decisions were issued, and many of those were subsequently reversed by the Trump board. In fact, I have said that I sometimes feel a bit canceled! But I spent eight years in the minority during the George W. Bush years, and I have to say that perhaps what I am most attached to, and may be most proud of, is the body of dissenting opinions because they can still migrate to a majority position at some time in the future.

But I also want to comment, Josh, on your remarks about democracy because that was the theme 100 years ago, and it was also a Louis Brandeis notion that you couldn't really have political democracy without democracy in the workplace. And the labor question, which people say has arisen again in this new Gilded Era, was the acknowledgment that there was a gap between the promise of democracy or the American dream and the grim realities of life for so many people, especially for so many working people. So, I think these issues are as real today as they were 100 years ago, particularly as we are seeing the erosion of some democratic institutions. I think these issues are as important as they have ever been, and I think workers do realize that.

A couple of bills that have been introduced in Congress would mandate worker representation on corporate boards. As I said earlier, this is part of the German model of co-determination that has fascinated some people. I don't think we can simply import the German model into our society. But I think, and this is what I tried to say at the beginning, I think we need to be open minded to new approaches, not in place of traditional unionism or collective bargaining by any means, but to perhaps open up the dialogue, to engage more about other avenues.

One of the things I think is saddest is the lack of any kind of meaningful dialogue between labor, business, and government, like what exists in Europe—the so-called social dialogue involving the social partners. That dialogue is sorely missing in our climate. I would like to hope that this historic moment will lead to some more opening up of that dialogue. Whether it's going to be a transformational moment in terms of profound change and improvements, I guess that remains to be seen.

It will require, I think, a lot of public attention, a lot of writing by you and your colleagues about these issues, and an opening up in the business world. It will take some courageous business leaders who are willing to say we need a new approach, that we can't just have this traditional knee-jerk anti-unionism. And I believe that the business world has to do that—that they, at their peril, ignore the alienation, the frustration, the voicelessness, the powerlessness of workers. We are seeing the results of that in many respects right now.

II. The Impact of Age, Gender, and Race on Retirement Decisions During the Pandemic Recession

The Duration of US Joblessness and the Great Recession

FERIDOON KOOHI-KAMALI
AIDA FARMAND
JOSÉ PEDRO BASTOS NEVES
The New School

Abstract

We develop models of duration of unemployment for the United States and apply them to the 1996–2020 biennial Displaced Worker Survey. Our approach has several notable features relevant to an analysis of unemployment during a deep, persistent recession. We estimate single-spell proportional hazard exponential and Weibull models of transitional probability to full employment, and we test the outcome for robustness to functional forms by distributional free estimation. We test for the impact of unobserved heterogeneity by modeling recession as a gamma-distributed unobservable frailty function. We also estimate a non/semiparametric proportional hazard Cox model of competing risks with several destinations out of unemployment: transition to full-time employment, part-time employment, and inactive job-search status. Our single-destination estimates demonstrate significant differences in the covariate estimates as a result of different base hazards defined for those during the Great Recession and those outside that period. The estimates show gender is the single most important determinant of the duration of unemployment, with men's hazard rate of (re)employment to full-time jobs 7 to 15 percent lower than women's. When interacted with unemployment insurance, being female also accounts for a 10 percent higher hazard of transition to full-time employment. Our competing risks regression results indicate that Black workers have 28 and 27 percent lower rates of exit to a full-time and part-time employment, respectively, following job displacement.

Introduction

Understanding how much time individuals spend in unemployment, how this duration changes over the business cycle and how the length of unemployment varies across individuals are important aspects of job loss because the welfare of the unemployed is closely related to the time they spend without a job. Moreover, the length of unemployment plays a critical role in economic theories of job searching. Three aspects on the duration of unemployment stand out in most unemployment survival studies: first, with regard to provision of social security programs such as unemployment benefits—for example Katz (1986); second, with regard to gender as the main determinant of exit from the labor market—for example Mussida (2007); and third, on the impact of recession on the duration of unemployment—for instance Pissarides (2013). In this study, we develop models of duration of unemployment for the United States covering the Great Recession period that address the above aspects of unemployment with several notable features relevant to an analysis of unemployment during a deep, persistent recession.

We estimated probability of exit to full employment by first examining two models that allow the exit time to be independent of and dependent on the unemployment period, and then we tested the outcome for impact of unobserved heterogeneity from effects such as periods of expansion and contraction, and robustness to proportional hazard parametric functional forms. We also estimated a multiple-exit model of unemployment duration with distinct exit destinations out of unemployment for full-time employment, part-time employment, and discouraged-worker status by a non/semiparametric competing risks model. We focused on gender and race of workers as the principal determinant of difference in return to employment and how that is impacted by unemployment insurance and applied the above approach to the 1996–2020 biennial Displaced Worker Survey (DWS).

Literature Review

Studies on unemployment duration are concerned with factors that affect an individual's probability of leaving unemployment, and these can be categorized by a few areas frequently highlighted in the literature. An understanding of these factors has widespread implications for designing unemployment public policies. One area studied extensively is the impact of gender on the duration of unemployment. Until the 1990, men in OECD countries were expected to have lengthier unemployment spells than women, but the difference disappeared for subsequent periods (Albanesi and Sahin 2018). Studies focusing on the 2000s found that in some countries, women are likely to have a lengthier unemployment spell than men (Azmat, Güell, and Manning 2006; Mussida 2007; Tansel and Taşçi 2010). The common explanations to this pattern are high reservation wages for women (possibly due to household activities) and fewer job opportunities. Tansel and Taşçi (2010), studying the Turkish economy, found that the probability of being unemployed after 12 months was 90 percent for women and only 70 percent for men. In particular, gender appears to be an important variable determining exit from the labor force following unemployment (Mussida 2007).

The question is which factors lengthen the searching period and which do not. In this respect, family structures are often considered a determinant factor. The literature has found meaningful and positive relationships for marital status and children's ages (Mussida 2007; Ollikainen 2003). Personal characteristics, such as age, citizenship, and education, were also found to be relevant. Foreign citizenship and age increase the unemployment period (Mussida 2007; Ollikainen 2003), while level of education is likely to decrease it (Ollikainen 2003). The impact of family composition was more recently outlined by Kuhlenkasper and Steinhardt (2011). The results show that the likelihood of returning to employment for German women is reduced by the presence of young children and older relatives in the household. According to Imbens and Lynch (2006), the number of children living at home has a negative effect on the chances of women finding a job. Tansel and Taşçi (2010) found that married men have a higher opportunity cost of unemployment and thus search more intensively.

Another topic often considered is the impact of education on the duration of unemployment (Boršič and Kavkler 2009; Dănăcică and Babucea 2007; Landmesser 2011; Tansel and Taşçi 2010). A positive impact of education on unemployment duration is not self-evident because while highly educated workers may easily find jobs in an increasingly technology-based economy, low-skill jobs are usually associated with a high rate of turnover (and thus a smaller unemployment period). According to Boršič and Kavkler (2009), people with professional college degrees or bachelor's degrees are better off than unemployed people with a master's degree. Dănăcică and Babucea (2007) suggest a negative relationship between unemployment and education. Education can matter not only in level but in quality. In other words, the kind of education, whether technical, theoretical, or vocational, also affects the likelihood of finding a job after unemployment. According to Landmesser (2011), vocational training is important in exiting unemployment. Tansel and Taşçi (2010) found that vocational high school graduates have higher unemployment exit probabilities compared to traditional high school graduates.

Furthermore, one of the mostly widely studied determinant of unemployment duration is the impact of social security programs such as unemployment benefits on the job searching process (Heath and Swann 1999; Katz 1986). According to Nickell (1979), the impact of income replacement ratio (ratio of income to benefits) was lower for those in long-term unemployment, while a high replacement ratio was associated with a longer duration of unemployment (Lancaster 1979). Grubb (2011) and Barro (2010) found a strong and considerable influence of unemployment insurance benefits in increasing the unemployment rate. Contrary to results from Grubb (2011) and Barro (2010), Rothstein (2011) found that unemployment insurance extensions had significant but small negative effects on the probability that the eligible unemployed would exit unemployment, with the effects concentrated on the long-term unemployed. The estimates imply that unemployment benefit extensions raised the unemployment rate in early 2011 by only about 0.1 to 0.5 percentage points, much less than was implied by earlier analyses.

The same small but significance influence of extended benefits was found by Farber and Valletta (2015) while studying the effects of the extended unemployment benefits on unemployment duration during the Great Recession. The small influence found is attributable to the recipients exiting the labor force—and not to the reduction of job finding. However, they argue that this may have long-term consequences for the natural rate of unemployment.

Finally, the literature has also considered the effects of recession periods in shaping the unemployment period. Researchers have paid a great attention to the financial crisis of 2008–2009 and its aftermath, known as the Great Recession. The reason is that when compared to other economic recessions, the Great Recession stands out as having the deepest and most widespread impact on the labor market (Pissarides 2013). Recession periods, in fact, can not only change unemployment rates but also the patterns of unemployment. In other words, it raises the question of how different factors conditioning the length of a joblessness spell vary during a recession period. Focusing on heterogeneous effects of gender, Sahin, Song, and Hobijn (2010) investigated the specific effect of recession on the gender gap in the unemployment rate. They found that women were hit less hard—thus narrowing the gender gap—mainly because men occupy most jobs in highly procyclical sectors and because men continued to search for jobs instead of quitting the labor force. Several studies have shown that the gender gap decreased during the recession. De la Rica and Rebollo-Sanz (2017), studying the Spanish labor market, found that recovery-brought jobs were more equitably distributed. Kroft et al. (2016) studied the effects of the Great Recession on long-term unemployment. Importantly, they found that the exit from unemployment is less likely after longer periods of unemployment. Barnichon and Figura (2013) used a counter-factual analysis to address the effects of recession: they fitted the model for a long period before the recession (1967–2012) and used the calibrated parameters to fit the recession data. The discrepancies that arose were attributed to the recession effect.

In this study, we developed models of duration of unemployment for the United States covering the Great Recession period to examine the factors highlighted in the literature that have several notable features relevant to an analysis of unemployment during a deep, persistent recession. We estimated probability of exit to full employment by first examining two models that allow the exit time to be independent of and dependent on the period of unemployment, and then we tested the outcome for impact of unobserved heterogeneity from effects such as periods of expansion and contraction. We employed a frailty gamma function borrowed from biostatistical treatment of the unobservable. We also re-estimated our regression results by a Cox nonsemiparametric hazard specification and interpreted these as two alternative methods of accounting for the impact of the recession on slope estimates. Since evidence suggests that inactive job seekers, as opposed to those out of the labor market, account for a large share of the unemployment, we also estimated a multiple-exit model of unemployment duration with three distinct exit destinations out of unemployment for full-time and part-time employment and for discouraged-worker status with a non/semiparametric competing risks model. Following the empirical evidence from the literature, we focused on gender of workers as the principal determinant of difference in return to employment and applied the above approach to the Current Population Survey's Displaced Worker Supplements (1996–2020). However, the multiple-exit results are not yet fully examined for recession frailty with alternative modeling methods and must be treated with caution.

More specifically, we contribute to the duration of unemployment literature by providing an intersectional analysis of unemployment duration, in which we examine the differences in duration of unemployment and exit probabilities across groups of white men, white women, nonwhite men, and nonwhite women. Our intersectional analysis reveals a more nuanced picture of the duration of unemployment. Intersectional analysis also allows us to capture a clearer picture of the racialized impact of unemployment during a period of severe contraction of the US economy. In this sense, our contributions are fourfold:

- to investigate the dependency of the prospect of getting a job in the US economy conditional on time spent unemployed
- to determine how the probabilities of exiting unemployment was impacted by the Great Recession

- to determine the distribution of multiple transitional routes out of employment duration
- to determine the extent to which socio-demographic covariates influence the length of stay in unemployment

Data

The data for this paper were obtained from the 1996–2020 biennial DWS, published as a supplement to the January CPS. All participants older than 20 years of age were asked whether they lost a job within the three years prior to the survey date. Those responding positively to the question were then asked a series of questions about the lost job and the period of subsequent unemployment.

These data allow us to analyze a variety of labor market outcomes following displacement, including earning losses, the duration of unemployment, and unemployment benefit–claiming behavior. Displacement is defined as involuntary separation based on operating decisions of the employer, such as a plant closing, an employer going out of business, or a layoff from which the worker was not recalled.

Calculation of the *disregarded* element of wages—that is, earning after adjustment for taxation purposes for lost earning due to periods of unemployment—is an important component of wages in the data sets. However, there is no uniform set of rules for disregarded earning because the conditions are determined at the state levels and the time-consuming task of obtaining separate amounts for each state proved beyond the limits of this study.

Econometric Models

Hazard Models

There are two main approaches in estimating the impact of exogenous variables on the duration of unemployment spells and the probability of re-employment. One is the structural form framework, and the other is the reduced form framework (Narendranathan and Nickell 1986). In this study, a reduced form framework was used because of its simplicity and the fact that it permits us to estimate the impact of exogenous variables without imposing too many distributional assumptions. The reduced form framework is concerned with the specification and estimation of the conditional probability of leaving unemployment. Survival analysis was used to model unemployment duration and to assess its dependence on explanatory variables. Search theory was then adopted to choose the covariates that must be included in the model and to interpret the estimated coefficients.

We denote the duration of an unemployment spell, the time at which unemployment comes to an end, by $T \geq 0$ and its cumulative distribution function among the population by $\boldsymbol{F(t)}$, where t denotes a particular value of T. We define the survivor function $\boldsymbol{S(.)}$as the complement function to the *cdf*—i.e., periods of time taken to come out of the initial state; that is,

$$S(t)=P[T>t] =1 - F(t)), t \geq 0. \quad (1)$$

(1) is the probability of surviving unemployment status past time t. Then with a change in time $t + h$ for a "small" $h > 0$, the probability of exiting the initial state in the time interval $[t, t+h)$ is given by

$$P\,(t \leq T < t+h | T \geq t) \quad (2)$$

For each $\boldsymbol{t}$, we define the hazard function $\lambda(t)$ as the probability of instantaneous rate of exit per unit of time, given a small change in t by h. The hazard function provides an approximation for the conditional probability (2) by

$$P\,(t \leq T < t+h \mid T \geq t) \approx \lambda(t)h \quad (3)$$

Comparing Single Risk Models: PH Exponential and Weibull

The central issue in survival analysis is the shape of the hazard function, and the role of (3) in survival or duration analysis is to defined by a density function for the distribution for t. In this study, we examined three such functions. The first two often provide benchmarks for comparison with other hazard distributions. The simplest is the exponential hazard function because it has a constant duration parameter; that is,

$$\lambda(t) = \lambda \text{, for all } t \geq o \qquad (4)$$

(4) makes the hazard function memoryless; the probability of exit in the duration period does not depend on how much time has been spent in the initial state. Hence, (1) becomes

$$S(t)=exp\left(-\int_0^t \lambda(t)dt\right) = exp(-\lambda(t)) \qquad (5)$$

The exponential function is a one-parameter distribution, and that makes it too restrictive in many applications. We also employ the Weibull distribution as a second, more flexible and duration-dependent alternative based on the generalization of the exponential distribution. If T has a Weibull distribution, its *cdf* is presented by $F(t)=1-exp(-\gamma t^{\alpha})$; its density is given by

$$f(t) = \gamma\alpha t^{\alpha-1}exp(-\gamma t^{\alpha})$$

and its hazard function, by

$$\lambda(t) = \frac{f(t)}{S(t)} = \gamma\alpha t^{\alpha-1} \qquad (6)$$

The values of the nonnegative parameters γ and α determine the exact shape of the Weibull distribution. The Weibull distribution reduces to the exponential distribution as its special case when $\alpha = 1$, demonstrating the Weibull as a more flexible generalization of the exponential. If we know the hazard to be increasing ($\alpha > 1$) or decreasing ($\alpha < 1$), then the Weibull distribution is suitable for capturing duration dependence.

The models specify the conditional hazard rate as the product of two separate functions:

$$\lambda(t|\text{x}) = \lambda_0(t,\alpha)\ \varphi(\text{x}\beta), \qquad (7)$$

where the hazard rate, $\lambda_0(t,\alpha)$, is a function of t alone, while $\varphi(x\beta)$ is a function of x alone and usually is specified by $\varphi(x\beta) = \exp(x'\beta)$. Hence, both the exponential and Weibull distributions can be defined as the proportional hazard (PH) models since their hazards are, respectively, $\exp(x'\beta)$ and $\exp(x'\beta)\alpha t^{\alpha-1}$. We note that since the baseline is scaled by the vector of covariates, if all covariates are zero, then $\lambda_0(t,\alpha)\beta_0$, where β_0, the intercept constant, becomes a part of the baseline.

Recession and Functional Form Effects

Estimation based on exponential and Weibull densities is vulnerable to misspecification errors for two reasons. First, we need to account for the impact of the various sources of unobservable heterogeneity on the estimation results—most importantly, the impact of the recession and expansion periods on employment. Second, any functional form misspecification can also lead to inconsistent estimates. We next address each one of these two issues in turn.

To address heterogeneity, we employ a mixture survival model that has unobserved heterogeneity explicitly entered into the hazard function. The mixture function is based on the following assumptions: (1) the heterogeneity is independent of the observed covariates, starting and censoring times, (2) the distribution of heterogeneity variable is known, and

(3) heterogeneity enters the hazard function multiplicatively. We employ a Weibull hazard function conditional on observed covariates x_i and an unobserved heterogeneity v_i by

$$\lambda(t|x_i, v_i) = v_i exp(x_i\beta)\alpha t^{\alpha-1}, \quad (8)$$

where $x_{i1} \equiv 1$ for the intercept, and the multiplicative hazard is $v_i > 0$; the identification of β and α requires conditioning on normalization for the distribution of frailty v_i. The applications usually adopt a gamma-distributed heterogeneity function, $v_i \sim$ gamma (δ, δ), and $E(v_i) = 1$ and Var = 1/δ. This method is one way to capture the unobservable effects of a recessionary period on estimations obtained, and it is commonly employed in biostatistics—for example, Hongaard (1986). In practice, it may be hard to distinguish between the duration dependence parameter α and heterogeneity v_i without repeated spell observations. However, if the test of Ho**:** α = 1 is confirmed, then that outcome can remove the effects of the Weibull flexibility parameter and hence isolate that of heterogeneity, particularly that for a recession, on the duration function. But the assumption of α = 1 may not be confirmed, and in any case, unobserved heterogeneity is still vulnerable to functional form misspecification error. Hence, we also employ an alternative distribution-free method to measure the recession effect.

The PH models are known to provide consistent estimates of the regression parameter vector by nonsemiparametric methods without specification of the functional form for λ_0. We employ a PH Cox semiparametric survival function that offers robustness against (4) and (6) in order to estimate β, unlike the HB (7) and (8) models, without estimating the base hazard at the same time. The Cox model is semiparametric in that the hazard rate λ_0 drops out of the estimation of β as a consequence of the PH assumption, though the λ_0 estimate can be recovered once β is estimated. Hence, the Cox model with the exponential covariant specification leaves λ_0 in (7) unspecified, now denoted as $h_o(t)$:

$$\lambda(t|\mathrm{x}) = h_0(t)\ \exp(\mathrm{x}\beta) \quad (9)$$

Specifically, the estimates by this model are presented in Table 2. The Cox proportional hazards regression model states that the survival function for the *i*th person in the data is estimated by

- *β:* a vector of unknown parameters to be estimated from the data
- $h_0(t)$: the baseline hazard function at time *t*; that is, the hazard function when all predictors are equal to zero
- *x*: an independent vector of predictor variables; the list can be found in Table 1

Note that the effect of recession in both methods is defined with reference to the base hazard function, multiplicatively for unobserved heterogeneity, unspecified and removed semiparametrically from the regression equation based only on the individual covariates. That may still leave unaccounted the direct recession effects on the covariates. If there are doubts regarding the presence of such direct effects on the covariates, we can also move a covariate of interest inside the hazard function then re-estimate the model separately in terms of that covariate—that is, if necessary, we can define the hazard function to have the same shape for all individuals but separately estimated in terms of a critical variable (e.g., by gender or race). This method, however, becomes unyielding with more than one variable if several separate questions have to be estimated for several different hazard functions.

Estimation of the duration models is complicated by the presence of right-censoring of the length of possible incomplete spells of unemployment. The survival model data are typically right-censored from above. With right-censored data, we observe spells from time 0 until a censoring time *c*; some spells are completed by the time *c*, while all we know for the other spells is that they will end at some time in the interval (c, ∞) since, in such cases, we know only that the duration exceeded *t*. Hence, in application, we need to specify a hazard function that accounts for right-censoring. For the hazard function (7), the uncensored contribution to the log *ML* function and the Weibull survival function lead to the Weibull distributed log *ML* as

$$\ln \mathrm{L} = \sum_{i=1}^{N}[\delta_i\{x_i'\beta + \ln\alpha + (\alpha-1)\ln t_i \exp(x_i'\beta)t_i^{\alpha}\} - (1-\delta_i)\exp(x'\beta)t_i^{\alpha}] \tag{10}$$

The differentiation of (10) with respect to β and α leads to the first-order conditions that solve for their estimates.

Competing Risk Model

Competing risk is a class of survival models designed for multiple transitional exit destinations. McCall (1996) developed a model in which workers could be re-employed either in full-time or part-time jobs and, for each outcome, the considered factors were allowed to have varying effects. Exit from the labor force can also be thought of a potential unemployment outcome. After a period searching unsuccessfully for a job, the person may quit the job market altogether. The distinction between transition from employment to out of the labor market by discouraged workers is particularly important during a long recession. Analyzing the Great Recession, Elsby, Hobjin, and Sahin (2010) found the nonparticipation exit was particularly important after the recession, while the flow to employment was essentially flat. Moreover, unemployment benefits may affect the outcome by making it possible for the unemployed to search for job for a longer period than otherwise would be possible (Rothstein 2011). This case is an alternative outcome to a full- or part-time job because though still in the job market, the person ceases their job search. Farber and Valletta (2015) found a negative influence of unemployment benefit when the outcome is labor force exit, suggesting that those recipients stick to the job market, perhaps with a lower search level. Krueger, Cramer, and Cho (2014) suggest that the likelihood of exiting the workforce is highly influenced by the duration of the unemployment spell, with longer periods being associated with higher labor force withdrawals. It is important to investigate exit from the workforce because it may be associated with specific and particularly strong economic effects, such as hysteresis (DeLong et al. 2012), skill mismatches, and housing performance (Estevão and Tsounta 2011).

We expanded the above models to account for the probability survival when there are more than one exit destinations—i.e., moving from unemployment to either full-time or part-time employment. A survival function with this type of multivariant transitional probability will involve estimating a joint distribution of durations. We employed the competing risk model (CRM) to estimate a multiple hazards version of the single-spell model (10), where each exit destination provides one complete duration *m*, and *m* – 1 censored durations; thus, competing risks determine the destination state. We denote destination-specific covariates by $x_j (j = 1,2,\ldots,m)$, and note that only one duration, the shortest, is observed at the end of the spell, $\tau = min_j(t_j)$, $t_j>0$. If the risks are independent, then the multiple-spell survival function is

$$S_\tau(t) = \Pr[\tau > t] = \Pr[t_1 > t] \times \ldots \times \Pr[t_m > t].$$

Let $g_j(t)dt$ be the probability of risk *j* materializing over a small interval change by *dt*. Then the total hazard rate of all durations is

$$\lambda_\tau(t) \equiv -d/dt \ln S_\tau(t) = \sum_{j=1}^{m} g_j(t);$$

that is, the probability of exit remains the same regardless of whether *j* is one of the risks or the only risk.

Given hazard function independence, the PH Cox duration model provides probability estimates for the integrated hazard over different destinations by a PH model of the form

$$\lambda_j(t;x) = \lambda_{0j}(t)\exp[(x'(t)\beta_j], j=1,\ldots,m, \tag{11}$$

where both the base hazard and parameter are specific to *j*-type hazard $t_{j1} < \ldots < t_{jk_j}$; k_j denotes the ordered destination of type *j*. For instance, with $m = 2$, k_1 = full-time work and k_2 = part-time work.

Specifications of the Econometric Equations

Below we specify Equations (4) and (7) through (14) for the single-spell duration, and Equations (15) and (16) for the triple-spell competing risks Cox model.

Exponential and Weibull Models

$$\lambda(t|\mathrm{x}) = \lambda_0(t, \alpha)\ \varphi(\mathrm{x}\beta), \tag{12}$$

where β is the vector of unknown parameters to be estimated from the data, $\lambda_0(t, \alpha)$ = baseline hazard function at time t (that is, hazard function when all predictors are equal to zero), and x = independent predictor variables (see Table 1 for the list of the variables). For exponential and Weibull models, their hazards are, respectively, $\exp(\mathbf{x}'\beta)$ and $\exp(\mathbf{x}'\beta)\alpha t^{\alpha-1}$. Specification for our model and estimates are presented in Table 2.

Unobserved Heterogeneity Model

Effective estimation of the recession unobservable frailty effect requires a recession dummy inside the frailty gamma function in Equation (7):

$$V_{id} \sim gamma\ (\boldsymbol{\delta},\ d), \tag{13}$$

where d = 1 for the Great Recession, and zero otherwise. The frailty implementation with this procedure, however, proves difficult with a large sample size of this study, and we have to adopt a modification discussed below.

Cox Semiparametric Model

$$\lambda(t|\mathrm{x}) = h_0(t)\ \exp(\mathrm{x}\beta) \tag{14}$$

The Cox proportional hazards regression model states that the hazard rate for the ith person in the data is β is the vector of unknown parameters to be estimated from the data $h_0(t)$ = baseline hazard function at time t; that is, the hazard function when all predictors are equal to zero and x = independent predictor variables. The list of variables can be found in Table 1, and specifications for our model and estimates are presented in Table 2.

Competing Risk Model

The estimates in Table 3 (competing risk model) assume that the full-time and part-time re-employment hazards are independent and that the regressor coefficients are time constant. Moreover, it is assumed that the probability of being re-employed into a type w job in period k is conditional on remaining jobless for more than $k - 1$ periods, indicators for unemployment insurance receipt (UI), gender and their interactions, and a J-dimensional vector of other regressors **x**, satisfies the following:

$$P(T_w = k|UI, SEX, RACE, x, T > k-1) = 1 - \exp \\ \left(-exp(\alpha_{wk} + g_w(UI, race, gender) + \beta_w x)\right), \tag{15}$$

where w = full-time or part-time. Here, α_{wk} are the baseline hazard parameters for risk w, and β_w is a J-dimensional vector of parameters measuring the effect of the other regressors on the conditional probability of re-employment into a type w job.

Moreover, g_{wk} is assumed to satisfy

$$g_w(UI, race, gender) = \gamma_{1w}UI + \gamma_{2w}race + \gamma_{3w}gender + \gamma_{4w}genderUI + \gamma_{5w}\mathrm{raceUI},$$

where *genderUI* and *raceUI* denote the interaction of the gender with unemployment insurance receipt, and the interaction of race with unemployment insurance receipt, respectively; and γ_{iw} represents the effect of the ith regressor on risk w, $i = 1,\ldots,5$; w = full/part time.

Furthermore, we model unobserved heterogeneity as follows:

$$P(T_w = k|UI, SEX, RACE, x, \theta_w, T > k-1) = 1 - \exp(-exp(\alpha_{wk} + g_w(UI, race, gender) + \beta_w x)\theta_w) \quad (16)$$

The univariate random variable θ_w, which affects the conditional probability of re-employment into a type w job is assumed to be unobserved.

We present our competing risks results as provisional, and although they remain to be completed in combination with the unobserved recession frailty analysis, we feel the tentative results are suggestive and sufficiently interesting despite their limitation.

Results

Single Risk Model

The CPS allows us to distinguish among three different exits after a period of unemployment: partial job, full-time job, and exit from the labor market. While we keep track of alterative post-displacement outcomes through four binary variables (Censor 1, Censor 2, Censor 3, and Censor 4), our initial focus is on the transition to a full-time job. We leave part-time work and exit from the labor force transitions to the next section. Thus, a spell of unemployment is complete if person is re-employed at a full-time job. We analyzed the factors conditioning it through the variables specified in Table 1, which provides information about the specification of variables used in the analysis.

Although not exhaustive, the variables selected address the most important characteristics with which we are presently concerned.

Table 1. Summary of variables used in the analysis

Variable label	Variable description	Mean or share
Weeks Unemployed	Number of weeks unemployed	13.1 (mean)
Censor 1	1 if re-employed full time	0.62
Censor 2	1 if re-employed part time	0.24
Censor 3	1 if unemployed	0.07
Censor 4	1 if not in labor force	0.04
UI	1 if received unemployment benefits	0.45
ColDegree	1 if individual has a college degree	0.28
Black	1 if individual identifies as Black	0.09
Female	1 if individual identifies as male	0.43
Tenure	Length of time worked at lost job, in years	4.7 (mean)
Log Wages	Logarithm of weekly earnings at lost job	6.3 (mean)
Age	Age of individual	39.5 (mean)
Young Child	1 if individual has children under the age of 8	0.23
Married	1 if individual is married	0.56

Source: Authors' calculations using Current Population Survey Displaced Workers Survey (1996–2020).

Additionally, we considered several specifications to our regression models. We started with the PH versions of exponential and Weibull parametric functional forms. The results are reported in Table 2. Hence, this suggests the duration of an unemployment spell is not a critical influence on transition to a full-time job. For almost all variables, the results by exponential and Weibull hazard densities are very close to each other. However, the third rows for Cox regression estimates are critical to support the robustness of these estimates to the unobserved recession impact, and on the face of it, those too support the parametric estimates; apparently, they are not vulnerable to either functional form misspecification or to the impact of the Great Recession period.

Table 2. Unemployment duration: Estimated hazard ratios from three parametric models

	Exponential		Weibull		Cox PH	
	B	z	B	z	B	z
COLDEGREE	1.354646	11	1.252752	11.13	1.201624	10.01
MALE	0.8648666	–3.5	0.9151363	–2.77	0.9365942	–2.18
BLACK	0.7082782	–6.33	0.7502406	–6.44	0.7543529	–6.66
UI	0.6460056	–14.27	0.6849512	–16.9	0.6460975	–21.61
TENURE	0.9804468	–10.71	0.9851854	–10.34	0.9859965	–10.66
LOGWAGE	1.599015	26.65	1.459267	28.82	1.413475	28.87
OVER55	0.4068767	–26.44	0.4629416	–27.27	0.4824331	–27.71
GENUI	1.158473	3.24	1.130009	3.5	1.128492	3.77
RUI	1.13612	1.7	1.109842	1.7	1.123324	2.04
YOUNG CHILD	1.063411	1.66	1.060442	2.22	1.06584	2.67
MARRIED	1.409103	10.66	1.34676	12.18	1.330283	12.75
GENDER*CHILD	0.6634504	–7.5	0.6975587	–8.46	0.702525	–8.96
MARRIED*GENDER	0.5958051	–10.97	0.6435615	–12.05	0.6547247	–12.58

Notes: The variable *RUI* is defined as an interaction term between unemployment benefits and being Black. *GENUI* is defined as an interaction term between unemployment benefits and being female. *GENDER*CHILD* and *MARRIED*GENDER* are defined as interaction terms between female and a dummy variable for having children and being female and a dummy variable for being married.

Source: Authors' calculations using Current Population Survey Displaced Workers Survey (1996–2020).

In all three specifications, gender is the single most important factor affecting the length of an unemployment spell. Men were found to take 7 to 15 percent more time before finding a new full-time job. This result is in line with findings for other countries (Mussida 2007; Ollikainen 2003). However, the fact that we are considering long-term data (since the 1996) does not allow us to say much about the shrinkage of this factor over time—as reported, for instance, in Albanesi and Sahin (2018). On the opposite side, we found that unemployment insurance diminishes the hazard of exiting the unemployment to 64 percent. The literature has for a long time considered this result as a shift in worker incentives. Our result so far then is apparently in line not only with classical studies on this question (Katz 1986) but also with recent research (Barro 2010; Grubb 2011). However, since this variable is most sensitive to parametric form, the Cox model produces a more significant result than the exponential one. Taken together with gender and race, however, the unemployment income variable exhibits a weaker influence. Interacting unemployment income with being Black or female (variable RUI and GENUI, respectively) increases the hazard functions to 73 and 74 percent, respectively.

The DWS does not collect information about unemployment income benefit eligibility or the amount of benefits that displaced workers receive. Unemployment benefit eligibility depends on both the earnings history of an individual and the reason for job loss. No individual in our sample would be disqualified for unemployment income benefits based on why they lost their job. Using the earnings information in the DWS, it would difficult to accurately determine whether a displaced worker satisfies the earnings requirements for unemployment insurance benefit eligibility.

Moreover, not all individuals eligible for unemployment insurance benefits apply to receive them. Elements that might impact workers' take-up decision need to be accounted for in the estimations since the unobservables determining unemployment insurance receipt may be correlated with other covariates, leading to selectivity bias.

Numerous factors impact unemployment insurance take-up decisions. An individual's contribution to total household income has the most significant effect on that decision. Other income variables appear to be less impactful. Total monthly household income matters, with those in the middle range (monthly income between $2,000 and $10,000) significantly more likely to take up unemployment insurance benefits relative to those with the lowest total monthly income (less than $2,000). The highest income category (more than $10,000) is not significantly different from the lowest group in the take-up decision. Additionally, the lowest income group may apply for other programs, such as food stamps or other welfare programs. In those cases, the household may not expect to be eligible for unemployment insurance benefits even when they are (Auray, Fuller, and Lepage-Saucier 2018). Given these challenges regarding take-up decisions, our estimates might underestimate the true impact of unemployment insurance on the duration of unemployment.

Recession

Recession periods have a strong impact on the job market, affecting both the ability of the economy in job demand and the behavior of workers on the supply side. Consequently, the covariates studied in this work are likely to have considerably different values outside a recession period. We aim at capturing such differences by fitting the unobserved heterogeneity with a gamma frailty function. Doing so allows our base hazard function to vary multiplicatively with the unobservable effects, principally for the Great Recession, thus capturing the recession effect. Additionally, we performed the same task but restricted the frailty function to estimate two separate survival equations—one for recession years and another for nonrecession years. The attempt to estimate the model with a frailty function based on a single dummy does not easily converge; hence, we adopted an alternative of treating recession/expansion as an ancillary variable. The results are reported in Table 3 below, where we considered workers who had held one (full-time) job since displacement.

Our results indicate that recession periods significantly alter most coefficients. Even the more general estimates with a nonspecific unobservable display notable differences for based on the PH Weibull distribution between the middle columns of Table 2 and the first columns of Table 3. In general, the impacts are much smaller when we controlled for the recession. Such differences became sharp when we compared the results restricted to the recession years (the last two columns) against those restricted for nonrecession periods (middle two columns). Increases in the chances of finding a job occurred with highly educated (0.42, from 0.28), male (0.42, from 0.30), and married (0.26 from 0.16) workers. The education effect is as expected, inasmuch as the positions that are eliminated during a recession are usually low-skilled ones, affecting disproportionately the low-skill workforce. The result for gender aligns with the findings of Azmat, Güell, and Manning (2006), which pointed to the recession as an important factor for closing the gender gap in unemployment rates. A lower transition hazard for men, as in our results, implies a decrease in the gender gap in unemployment rates.

Conversely, for other variables, recession had the opposite effect—namely, decreasing the hazard. That happened with the *Black* variable (insignificant for recession and –0.18 otherwise), *UI* (insignificant for recession and –0.16 otherwise), and the *over 55* dummy (–0.48 against –0.36 in nonrecession years). We noted that our estimate for unemployment insurance for the Great Recession period observations was no longer significant at 5 percent, in contrast to our earlier result and in contrast with Grubb (2011) and Barro (2010), as we find that benefits did not significantly impact the transition-to-employment likelihood during recession periods. Our results are more in line with Rothstein's (2011) findings that unemployment insurance extensions had small negative effects on the probability that the eligible unemployed would exit unemployment.

Table 3. Unemployment duration: Estimated hazard coefficients considering heterogeneity and recession effects

	Gamma frailty		Gamma frailty for nonrecession years		Gamma frailty for recession years	
	B	z	B	z	B	z
COLDEGREE	0.28	4.37	0.28	3.07	0.42	2.53
MALE	0.29	4.90	0.30	3.45	0.42	2.75
BLACK	−0.14	−2.51	−0.18	−2.29	−0.07*	−0.58
UI	−0.15	−3.91	−0.16	−2.94	−0.17*	−1.93
TENURE	0.01	3.36	0.01	2.48	0.01	1.97
LOGWAGE	0.11	4.02	0.11	2.86	0.16	2.22
OVER55	−0.36	−5.06	−0.36	−3.62	−0.48	−2.90
YOUNG CHILD	−0.04*	−1.49	−0.04*	−1.18	−0.04*	−0.56
MARRIED	0.17	5.00	0.16	3.75	0.26	2.61
RUI	−0.08*	−1.18	−0.03*	−0.31	−0.23*	−1.35
GENUI	0.17	1.70	0.15*	1.17	0.28*	1.15

Notes: The variable *RUI* is defined as an interaction term between unemployment benefits and race. *GENUI* is defined as an interaction term between unemployment benefits and gender. * Indicates non-significant coefficients.

Source: Authors' calculations using Current Population Survey Displaced Workers Survey (1996–2020).

As for age, we found a disproportionate burden of recession on older workers. Negative coefficients indicate decreased hazard and increased survival times, which translates to a prolonged spell of unemployment for older workers compared to younger workers in nonrecession years and especially in times of economic downturn.

Multiple-Risk Model

Next, we considered alternative post-displacement outcomes. That means the individuals may transition not only to full-time jobs (Risk 1) but also to part-time jobs (Risk 2). Risk 3 is to remain unemployed, which is treated in our model as a right-censoring case. The last outcome (Risk 4) is to leave the labor market altogether. Figure 1 compares the cumulative hazard estimates for each transitional destination for the PH Cox competing risks model; note that the part-time destination curve displays a steep rise compared to the relatively flat flow of return to work.

Figure 1. Baseline cumulative hazard functions

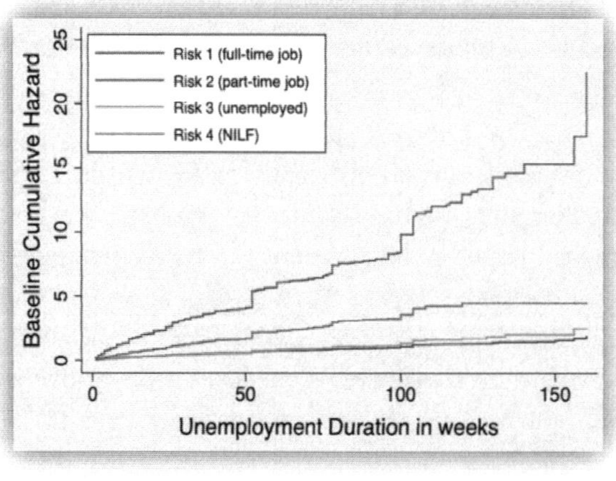

Black workers have 28 and 27 percent lower rates of exit to a job (full-time and part-time, respectively), while having a higher probability of remaining in unemployment (35 percent). Men also have a lower exit to (re)employment rate by 7 percent to a full-time job and 15 percent to a part-time job.

Interaction terms with having a child and marital status shed light on the gender role. For full-employment outcome, having a young child lowered the probability by 34 percent. The risk of exiting the labor force after displacement was also higher for women with children. Possible explanations include uneven household responsibilities and imbalanced intra-household bargaining power that imposes unpaid care work on women (Folbre 2021), difficulties finding a suitable job because of gender discrimination, and possibly higher reservation wages resulting from mothers' preferences for spending time with children (Youderian 2014). Being married was also associated with an increased (28 percent) rate of transition to a full-time job. Older workers had a lower rate (–6 percent) of transitioning to full-time re(employment) as well.

Table 4. Competing risks estimates for multiple transitional probabilities

Covariates	Risk 1	Risk 2	Risk 3	Risk 4
COLDEGREE	0.184	0.112	–0.157	–0.111
	(0.018)	(0.031)	(0.021)	(0.029)
Male	–0.066	–0.154	0.193	–0.117
	(0.03)	(0.042)	(0.032)	(0.04)
Black	–0.282	–0.277	0.35	–0.146
	(0.042)	(0.056)	(0.034)	(0.049)
UI	–0.437	–0.569	0.189	–0.5
	(0.02)	(0.038)	(0.023)	(0.037)
Tenure	–0.014	–0.009	–0.023	0.013
	(0.001)	(0.002)	(0.001)	(0.001)
Log wage	0.346	–0.134	0.22	–0.118
	(0.012)	(0.017)	(0.012)	(0.014)
Black*UI	–0.729	–0.276	–0.225	0.734
	(0.026)	(0.037)	(0.023)	(0.026)
Female*UI	0.121	0.186	0.138	0.092
	(0.032)	(0.05)	(0.035)	(0.046)
Young child	0.116	0.097	–0.267	0.049
	(0.057)	(0.084)	(0.044)	(0.076)
Over 55	–0.064	0.041	–0.039	–0.247
	0.024	0.047	0.027	0.062
Married	0.285	–0.087	–0.14	–0.099
	(0.022)	(0.039)	(0.022)	(0.038)
Young child*female	–0.353	–0.121	–0.143	0.733
	(0.039)	(0.061)	(0.042)	(0.068)
Married*female	–0.424	0.057	–0.062	0.311
	(0.034)	(0.052)	(0.034)	(0.047)

Note: Standard errors in brackets.

Source: Authors' calculations using Current Population Survey Displaced Workers Survey (1996–2020).

Finally, our results also lead to some considerations on the unemployment benefit. Irrespective of the outcome, there was strong evidence that unemployment benefits decrease the probability of unemployment transition, rendering the spell longer. The strongest effect happens in the part-time transition, which can be explained in terms of the benefit allowing workers more job search time. However, in view of the unemployment insurance estimate given in the last columns of Table 3, this impact may be negligible in a deep recession. The regression coefficient for the interaction term of female and unemployment insurance was positive, meaning an increase of the exit to a job hazard for women, while the coefficient associated with the risk of leaving the labor force was negative—indicating increased labor market attachment in presence of unemployment insurance benefits.

We end this section by noting the relatively sharp contrast between the estimates from the single and multiple estimation Cox estimates of Tables 2 and 4. For instance, gender and unemployment insurance estimates are 0.86 (z = –3.5) and 0.64 (z = –14.2), respectively, in Table 2, but 0.06 (s.e. = 0.03) and –0.43 (s.e. = 0.03). We discuss this outcome in the next section.

Discussion

The study suggests that unemployment insurance benefits impact unemployment spells and exit destinations differently across gender and racial groups. We also presented evidence that such benefits do not prolong unemployment spells in recessions.

One issue not examined in sufficient depth is the effect of age on this outcome. A nearly 50 percent lower probability of leaving the unemployment status for workers over 55 years old from our single-spell models was very high, and chances were, in the prolonged period of the Great Recession, most of them might well have been permanently pushed out of the labor market and never able to return once the economy rebounded. Retirement of these older workers earlier than required for adequate retirement saving has serious implications for old-age poverty and wellness. In future research, we plan to address the impact of age as a recessionary duration of unemployment influence.

Our treatment of the recession relies on what is commonly employed in biostatistics, but there may be more effective alternatives available from a state–space econometric modeling that specify a model for recession/expansion states and estimate the system of equations subject to unobservable state specifications. We intend to explore this approach in future work.

However, an important caveat is the limitation of our multiple-exit probability estimates without accounting for the recession frailty impact. Despite their limitation, the multiple destination transitional estimates were suggestive. A comparison of a single spell and a multiple spell of transition to full-time status by the PH Cox model revealed notably different estimates; the multiple-exit model differed significantly not just in magnitudes of covariates effects but also in the direction of those effects. Until further research, our findings in this regard must be considered provisional.

Conclusion

Our single-destination estimates showed that gender was the single most important determinant of the duration of unemployment, with men's hazard rate of unemployment status exit 7 to 15 percent lower than that of women, while unemployment insurance lowers the probability of transition to full employment by 64 percent. However, gender also accounted for this small hazard transition—an interactive term of unemployment insurance and gender lowered that transition by a range of 5 to 10 percent. Age was the other major effect on the unemployment transition, with workers over 55 years of age experiencing 50 percent lower hazard transitional probability, while being married increased that transition by nearly 30 percent. We also attempted to control for the impact of recession in estimations; the outcome apparently indicated robust parametric estimates in that regard.

The estimates from our competing risks regression for multiple hazards by the Cox non/semiparametric model also provided evidence for the critical role played by gender. Our results also indicated that race increased the length of the unemployment spell substantially when the individual transitioned to full-time employment (–27 percent). Since the spells differ primarily by the gender effect, the evidence pointed to gender of the worker as a crucial factor in the decision to leave the labor force. Interaction with children and marital status also shed light on the gender role. For full-employment outcome, having a young child increased the probability by 12 percent. For part-time and exit of out of the labor force, however, there was no statistically meaningful relationship found. There was also strong evidence that unemployment benefits decreased the probability of the unemployment transition, rendering the duration longer. The strongest effect happened in the part-time transition.

Finally, we suggested that our control for unobserved heterogeneity and functional form misspecification could be interpreted as a control for the impact of the recession. We presented the impact of the Great Recession on our co-variate estimation with a frailty function for the unobservables, and with separate frailty-multiplied hazard function for recession and nonrecession observations separately; the results pointed to large impacts of the recession on the estimates obtained, especially for the sample of the Great Recession period observations. The task of extending this procedure to our multiple destination probability estimation remains an important future work.

Acknowledgments

We would like to thank Teresa Ghilarducci, Siavash Radpour, Barbara Schuster, Karen Smith, Monique Morrissey, and Truc Bui for their comments and suggestions.

References

Albanesi, Stefania, and Ayşegül Şahin. 2018. "The gender unemployment gap." *Review of Economic Dynamics* 30: 47–67.

Auray, S., D. L. Fuller, and N. Lepage-Saucier. 2018. "Why Do Half of Unemployment Benefits Go Unclaimed?" http://www.fullerecon.com/tuem_0218.pdf

Azmat, Ghazala, Maia Güell, and Alan Manning. 2006. "Gender gaps in unemployment rates in OECD countries." *Journal of Labor Economics* 24.1: 1–37.

Barnichon, Regis, and Andrew Figura. 2013. "Declining labor force attachment and downward trends in unemployment and participation." SSRN 2376368.

Barro, Robert. 2010. "The folly of subsidizing unemployment." *Wall Street Journal*, Aug. 30.

Boršič, Darja, and Alenka Kavkler. 2009. "Modeling unemployment duration in Slovenia using Cox regression models." *Transition Studies Review* 16.1: 145–156.

Dănăcică, Daniela, and Ana-Gabriela Babucea. 2007. "The role of education for the duration of unemployment in Gorj Country." *Studia Universitatis Babes Bolyai–Oeconomica* 1: 67–82.

De la Rica, Sara, and Yolanda F. Rebollo-Sanz. 2017. "Gender differentials in unemployment ins and outs during the great recession in Spain." *De Economist* 165.1, no. 4: 67–99.

DeLong, J. Bradford, Lawrence H. Summers, Martin Feldstein, and Valerie A. Ramey. 2012. "Fiscal policy in a depressed economy [with comments and discussion]." *Brookings Papers on Economic Activity* : 233–297.

Elsby, Michael W., Bart Hobijn, and Ayşegül Sahin. 2010. "The labor market in the Great Recession." No. w15979. National Bureau of Economic Research.

Estevão, Marcello M., and Tsounta, Evridiki. May 2011. "Has the Great Recession raised U.S. structural unemployment?" IMF Working Paper No. 11/105. SRN: https://ssrn.com/abstract=1847338

Farber, Henry S., and Robert G. Valletta. 2015. "Do extended unemployment benefits lengthen unemployment spells? Evidence from recent cycles in the US labor market." *Journal of Human Resources* 50, no. 4: 873–909.

Folbre, Nancy. 2021. "Gender inequality and bargaining in the U.S. labor market." Economic Policy Institute.

Grubb, David. June 14, 2011. "Assessing the impact of recent unemployment insurance extensions in the United States." Draft presented at IZA, Berlin 25.

Heath, Alexandra, and Troy Swann. February 1999. "A structural model of unemployment duration | RDP 1999–02: Reservation wages and the duration of unemployment." Reserve Bank of Australia Research Discussion Papers.

Hongaard, P. 1986. "Survival models for heterogeneous populations derived from stable distributions." *Biometrica* 73: 387–396.

Imbens, Guido W., and Lisa M. Lynch. 2006. "Re-employment probabilities over the business cycle." P*ortuguese Economic Journal* 5.2: 111–134.

Katz, Lawrence F. 1986. "Efficiency wage theories: A partial evaluation." *NBER Macroeconomics Annual* 1: 235–276.

Kroft, Kory, Fabian Lange, Matthew J. Notowidigdo, and Lawrence F. Katz. 2016. "Long-term unemployment and the Great Recession: The role of composition, duration dependence, and nonparticipation." *Journal of Labor Economics* 34, no. S1: S7–S54.

Krueger, Alan B., Judd Cramer, and David Cho. 2014. "Are the long-term unemployed on the margins of the labor market?" *Brookings Papers on Economic Activity 2014*, no. 1: 229–299.

Kuhlenkasper, Torben, and Max Friedrich Steinhardt. 2011. "Unemployment duration in Germany: A comprehensive study with dynamic hazard models and P-splines." HWWI Research Paper No. 111.

Lancaster, Tony. 1979. "Econometric methods for the duration of unemployment." *Econometrica*: 939–956.

Landmesser, Joanna. 2011. "The impact of vocational training on the unemployment duration." *International Advances in Economic Research* 17.1: 89–100.

McCall, Brian P. 1996. "Unemployment insurance rules, joblessness, and part-time work. *Econometrica* 64.3: 647–682.

Mussida, Chiara. 2007. "Unemployment duration and competing risks: A regional investigation." *DISCES—Quaderni del Dipartimento di Scienze Economiche e Sociali* 0749.

Narendranathan, Wiji, and Stephen Nickell. 1986. "Estimating the parameters of interest in a job search model." In *Unemployment, Search, and Labour Supply*. Cambridge: Cambridge University Press.

Nickell, Stephen. 1979. "Estimating the probability of leaving unemployment." *Econometrica*: 1249–1266.

Ollikainen, Virve. 2003. "The determinants of unemployment duration by gender in Finland." Discussion Papers, VATT Institute for Economic Research.

Pissarides, Christopher A. 2013. "Unemployment in the Great Recession." *Economica* 80.319: 385–403.

Rothstein, Jesse. 2011. "Unemployment insurance and job search in the Great Recession." No. w17534. National Bureau of Economic Research.

Sahin, Ayşegül, Joseph Song, and Bart Hobijn. 2010. "The unemployment gender gap during the 2007 recession." *Current Issues in Economics and Finance* 16, no. 2.

Tansel, Aysit, and H. Mehmet Taşçi. 2010. "Hazard analysis of unemployment duration by gender in a developing country: The case of Turkey." *Labour* 24, no. 4: 501–530.

Youderian Xiaoyan Chen. 2014. "The motherhood wage penalty and non-working women," *Economics Bulletin* 34, no. 2): 757–765.

Resistance in the Digital Workplace

DEBBIE GOLDMAN
Communication Workers of America

Victoria Kintzer sat at her desk in the Bell of Pennsylvania residential business office, heard the ring of the black rotary dial telephone on her desk, and before it could ring again, picked up the receiver. The customer wanted to place an order for telephone service and also had a question about her bill. Kintzer looked for the customer's record in the file cabinet next to her desk. The file cabinet, known as a tub, contained all the records of customers in the local telephone exchange for which she was responsible. If the customer lived in a nearby exchange, Kintzer got up to get the record from a co-worker's tub, stopping for a moment to chat, and then returned to her desk to service the customer. She wrote up the customer's request, putting her notes in the out box for the service order writer to type up for distribution to the proper department. Although her supervisor listened in on some of her calls to ensure conformance with the detailed Bell system methods and procedures, and she resented the rules that made her raise her hand to go to the bathroom, Kintzer had relative autonomy to use her skills, knowledge, and emotional intelligence to assist her customers. She took satisfaction in helping them, solving problems, making sure the job was done right from start to finish. She knew that she could turn to her union if she had problems on the job. This was the work life of the Bell telephone service representative in the monopoly era in the 1970s and early 1980s, before the introduction of digital technologies and market forces drove change in its customer service operations.

Fast forward to 1994. Kintzer still worked for Bell of Pennsylvania, but her job as a service representative had changed dramatically. It was a decade since the 1984 breakup of the Bell system that accelerated market competition in the telecommunications industry. Kintzer now sat in a cubicle with shoulder-high dividers separating her from her co-workers. She trained her eyes on her computer, with the automatic call distributor dropping one call after another into her headset with no time between calls. She took only sales calls; the automated system sent billing inquiries to a different call queue staffed by lower-paid collections representatives. The job had become quite complex, with various rates for different geographic areas, multiple products and services, and data entry and retrieval from several software systems. Kintzer entered her service orders directly into the computer, frequently overlapping while taking the next call. Electronic scripts guided her through conversations with customers designed to upsell additional products and services. Digital software tracked her every move in real time—how long she took on each call, the time between calls, adherence to her daily schedule, sales performance, the manner in which she moved through her scripts and databases, even how many keystrokes she took to perform a task. Supervisors evaluated her performance on all these measures and imposed discipline for failure to meet sales quotas and time measurement benchmarks. As a local union leader, Kintzer faced challenges protecting her members from the intensified monitoring, sales pressures, and work speedup. This was the work life of the service representatives in the former Bell telephone companies in an era of increased competition and fully automated workplace management systems.

Within the space of a decade, Bell of Pennsylvania, along with the other legacy Bell system companies, had revolutionized the work life of tens of thousands of service representatives who worked in the customer care call centers. As the corporate executives transformed their businesses from monopoly-era regulated public service organizations with guaranteed rates of return to competitive enterprises in pursuit of ever-higher profits, the managers of the customer service operations struggled to reduce costs and boost revenues while still providing quality service. These managers turned to Bell system engineers, programmers, and outside vendors to enhance the digital technologies that distributed the calls to the service representatives, developing sophisticated workforce management systems that enabled managers to intensify control and increase surveillance over the pace and manner of work. The resulting work organization of high demand with little control was a recipe for debilitating and demoralizing stress on the job (Karasek 1979).

How did the service representatives wrestle with their employers over the degradation of the working conditions and downward pressure on living standards in the call centers? What strategies did they adopt to push back against the enormous power of automated technology to control and surveil their work? How effective were those strategies and under what conditions? Did the service representative's unique role as the face of the company to the customer elevate her power, and if so, how and under what conditions? What difference did it make that this struggle took place in a unionized setting? How did the union fight for greater influence over work organization, technology deployment, and the labor process, areas traditionally classified as management rights not subject to mandatory negotiation with the union? And how did larger social, economic, and political forces shape the boundaries and outcomes of these battles between labor and capital in the electronically managed call center environment?

"Resistance in the Digital Workplace: Resistance in the Digital Workplace: Call Center Workers in Bell Telephone Companies, 1965–2005" (my dissertation, published in 2021, University of Maryland) addresses these questions through a historical analysis of the ways in which a predominantly female unionized workforce contested the transformation of the labor process and downward pressure on living standards and job security in the customer service operations of two legacy Bell system telecommunications companies, AT&T and Bell Atlantic (now called Verizon). The introduction of new digital technologies, the intensification of market competition facilitated by those technologies and neoliberal regulatory policies, the financial turn in capitalism, and corporate and managerial business choices drove monumental changes in the AT&T and Bell Atlantic call centers in the latter two decades of the 20th century. The customer service workers served as the shock troops (who often felt like cannon fodder) on the front lines of this tumultuous transition.

"Resistance in the Digital Workplace" argues that AT&T and Bell Atlantic customer service workers coalesced into a workforce of resistance in response to the degrading, stressful conditions in their call centers. They demanded that their union, the Communications Workers of America (CWA), fight on their behalf for humane, safe, secure good-paying jobs. Yet the very forces that were driving change in the call centers also shifted and frequently narrowed the terrain upon which these call center workers struggled with management to shape their conditions at work and undermined the power of the union. In their struggles with employers, the call center workers and the CWA challenged neoliberal policies amid both the digital revolution and the decline of unions. "Resistance in the Digital Workplace" argues that while technology reconfigures the economic terrain, it is human beings, with competing interests, resources, and power, who make the decisions about how to deploy technology and which institutional structures will frame the social impact of those technologies. Those decisions about how to configure technology are social and political choices, with outcomes determined by political struggles between management and labor, shaped by the institutional structures of the political economy in which they take place (Braverman 1974; Burroway 1979; Callaghan and Thompson 2001; Edwards 1979; Noble 1986; Resnikoff 2021) As Karl Marx wrote in 1982, "People make their own history, but they do not make it just as they please; they do not make it under circumstances chosen by themselves, but under circumstances directly found, given, and transmitted from the past" (Marx 1986).

The AT&T and Bell Atlantic customer service workers had one important resource that most US call center workers, indeed most service sector workers, do not have: they were represented by a union. The CWA had a mature bargaining relationship with the Bell companies dating back to the post-World War II period, giving them a vehicle for collective resistance not available to most call center employees who work in an almost exclusively nonunion sector. To be sure, individual call center workers have some degree of agency as individuals in pushing back against conditions in the digital workplace—they can game the system, they can quit—but without organization, they have limited power to impact fundamental issues of work organization, technology deployment, job security, and compensation (Doellgast 2012; Doellgast and O'Brady 2021; Taylor and Bain 1999).

The CWA-represented customer service workers struggled for power not only with their employers but also within their union. Although militant women operators built CWA, the predominantly male technicians were the dominant force within the union by the 1960s (Green 2001; Norwood 1990). With technology gradually decimating the operator workforce, and competition elevating the size and strategic importance of the customer sales and service operation, the service representatives now fought for greater voice within their union. Labor historians have catalogued the struggles of women unionists for power within their unions. Dorothy Sue Cobble's *Dishing It Out* study of waitress unions makes a powerful case that the craft- and gender-based structure of the Hotel and Restaurant Employees (HERE) union gave power, voice, and autonomy to female waitresses that women unionists in amalgamated industrial unions battled to achieve. My case studies complicate Cobble's argument, finding that while female-led customer service bargaining units excelled at crafting local solutions to workplace problems and developing strong leaders, it was necessary to mobilize the full power of their diversified union to make significant progress on their issues. This was a formidable task, but when successful, as during the Verizon strike in the year 2000, it proved critical to winning a pathbreaking stress relief package after 18 days on the picket lines. (For a discussion of women workers and their unions, see Barry 2007; Benson 1986; Boris and Klein 2012; Cobble 1991; Gabin 1990; Kessler-Harris 2007; Milkman 1985; and Nielsen 1982).

"Resistance in the Digital Workplace" focuses on the contest between labor and capital in a pivotal industry of the post-industrial economy—telecommunications—and within that sector, customer sales and service. The digital revolution coupled with the growth of the consumer economy transformed the ways in which people buy products and seek customer service and support in the US economy (Cohen 2004; Stevens 2014). Beginning in the 1980s and accelerating since then, call centers became the primary vehicle through which businesses (and many public agencies) interact with customers, clients, and citizens. An estimated 3.7 million employees in the United States work in almost 40,000 domestic call centers, representing about 2.5 percent of the US workforce (Contact Babel, no date; US Bureau of Labor Statistics 2021a, 2021b). The global call center industry employs millions more (Holman, Batt, and Holtgrewe 2007; Lee 2015; Padios 2012). According to one research report, the US domestic call center industry generated $118.3 billion in revenue in 2020; global call center industry revenue was $404.3 billion that same year (StrategyR Global Analysts 2022). In 2020, the five largest call center companies employed 875,000 workers in more than 900 global call centers (Concentrix 2020, Sitel 2021; Sykes 2020; Teleperformance 2020; TTEC Holdings 2020).

While call center employees work in virtually every sector of the US and global economy, "Resistance in the Digital Workplace" focuses on call center workers employed by two leading progenies of the Bell telephone system, AT&T and Bell Atlantic. Many of the forces driving change for service workers in the late 20th century come together in a study of the telecommunications industry and the Bell corporate family: the impact of digital technologies on work organization and the labor process; the transformative impact of new technologies on market structure, competition, and consumer services; the ascendance of neoliberal free market ideology over the New Deal Order regulatory framework; the declining power of unions; and the role of emotional labor in the female-dominated service economy.

The Bell system provides an excellent case study to unravel the multiple forces driving change and setting the bounds for organized worker resistance in the customer service workplaces. Two distinct regulatory events accelerated market competition in telecommunications markets. The 1984 AT&T divestiture broke up the integrated Bell system monopoly. To resolve a longstanding anti-trust suit, the AT&T consent decree limited the company to long-distance service, equipment manufacturing and installation, and research in the world-renowned Bell Labs, while requiring it to spin off the local networks to seven regional Bell operating companies (RBOCs), which included Bell Atlantic. The RBOCs were limited to provision of local telephone service (Temin 1987). Bell system employees who weathered the 1984 dissolution of the AT&T monopoly talk about work life "before" and "after" divestiture. "Before" were the good old days; "after" led to massive job cuts, insecurity, and weakening of union power. Twelve years after divestiture, Congress passed the Telecommunications Act of 1996, which opened all telecommunications markets to competition and accelerated a dizzying wave of mergers, acquisitions, and cost-cutting pressures, including outsourcing of call center work. These two government policy changes serve as markers for transformation in market structure and competitive conditions that in turn shifted the terrain upon which the union and management contested for control. Union representation in the telecommunications industry declined during this period from about 60 percent in the early 1980s to 22 percent in 2005, the concluding year of this study. Today it stands at 13 percent (Figure 1; Hirsch and MacPherson 2005; Schmitt and Kandra 2020; US Bureau of Labor Statistics 2021a, 2021b).

Figure 1. Telecommunications employment and union density, 1983–2020

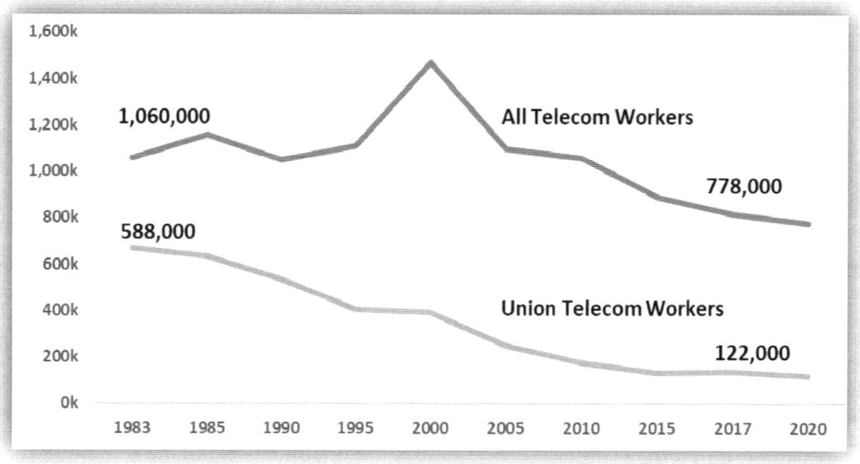

Source: Hirsch/Macpherson database (1983–2018); US Bureau of Labor Statistics (2019–2020.

Customer service union leaders and their members mobilized to promote an alternative vision of the customer service workplace. That vision emphasized quality customer service delivered by highly skilled professionals working in an environment that valued their judgment, experience, and skills to respond to customer needs. CWA customer service members and their leaders mobilized to fight for working conditions in the call centers that emphasized quality rather than low cost, based on the belief that the benefits in customer satisfaction and loyalty, as well as reduced costs from lower rates of turnover, absenteeism, and stress-related illness, could support higher wages, benefits, and investment in training. They articulated a program that would use the enormous power of the information generated by digital technologies as tools to improve worker performance and customer car—rather than for monitoring, speedup, and delivery of scripts for customer interactions.

Indeed, the union won important victories for its members: curbing secret supervisory monitoring, limiting discipline for failure to meet sales quotas and time measurement benchmarks, raising the wages of service representatives, fighting the functionalization and downgrading of nonsales positions, bringing back some outsourced work, and gaining access to the jobs selling and servicing what were then emerging new products including Internet (and at AT&T, wireless) services. But the union and its customer service members faced formidable challenges as their employers reorganized their businesses many times over to compete against nonunion companies with lower labor costs and more favorable regulatory treatment, while at the same time meeting the demands of capital owners to deliver ever-higher returns on their investments. Weak US labor laws provided little assistance to nonunion workers seeking union representation. The victories of the CWA call center workers frame the boundaries of organized worker power in contesting managerial control in the automated workplace in the context of the neoliberal political economy and financial turn of capitalism. The contests of this pivotal unionized workforce, operating in one of the most dynamic and important sectors of the US and global economy, highlight the opportunities, challenges, and constraints US service workers face in their struggles for power in the post-industrial service economy.

The digital transformation impacts virtually every workplace in today's highly automated post-industrial economy. From the Amazon warehouses to UPS drivers to modern manufacturing systems and hotel housekeepers, employers deploy the vast capabilities of information technology to regulate, track, and control the pace and the manner in which workers do their jobs. These electronic systems give employers tools that Frederick W. Taylor only dreamed of when he pioneered his system of scientific management in the factories of the early 20th century (Taylor 1911). Not only do they give management the ability to control workers' time and the labor process far beyond the machine pacing of the assembly line, but they also enable employers to break up internal labor markets and to outsource production and service functions to third parties across the globe (Weill 2014). In the call center, automated systems add a third dimension because they mediate the interaction between workers and customers. The same digital technologies that employers use to control the labor process are used by technology companies to track and monetize our lives as we move through websites, emails, texts, and social media posts. Business historian Shoshanna Zuboff calls this the Age of Surveillance Capitalism (Zuboff 2019). This study of the ways in which organized workers mobilized to resist dehumanizing labor in digital workplaces, and the constraints they encountered in the context of a neoliberal political economy that provides few guardrails against the ravages of financial capitalism, has wider implications as we re-examine the regulatory regime that has made this possible.

Labor historian Richard Edwards aptly describes the workplace as contested terrain, an arena for struggle between management and labor for job control and allocation of a firm's resources. In my study, I reinvigorate Edwards' theoretical framework, which describes a dialectical process in which workers' reactions to new technologies and work organization create a dynamic response and counter-response in a struggle over technology and job design. The contest between labor and capital takes place within a specific historical context, shaped by the degree of competition, size of the corporation, extent of union organization, level of worker class consciousness, impact of government policies, and speed of technological change (Edwards 1979).

The specific historical context for the labor–management struggle in the Bell system call centers is the ascent of the neoliberal political economy and the financial turn of capitalism and management. Beginning in the 1970s, in response to the economic challenges of global competition in basic industries, oil crisis, inflation, and unemployment, policy makers abandoned the New Deal Order's regulatory policies over finance, labor markets, industrial structure, and social welfare. The neoliberal faith in free markets weakened government institutions that had served for 40 years as a counterbalance to the power of private capital in the economy and workplace. Wealth and power flowed upward in the neoliberal political economy, one characterized by relaxed government of basic industries, finance, and labor markets; cutbacks in government spending on social welfare; tax cuts for the wealthy and corporations; privatization of public services; the marginalization of collective bargaining; and the casualization of jobs (Davis 2009; Epstein 2005; Gerstle 2022; Kotz 2017; Krippner 2011; Lazonick and Shin 2020).

The rise of the neoliberal order facilitated and accelerated the financial turn in capitalism and managerial strategies. During the years covered by this monograph, the former Bell companies transformed themselves from a Chandlerian model of managerial capitalism to a financial model. This transition, driven by capital's laser focus on boosting shareholder value, transformed the managerial function from one that coordinated vertically integrated activities within a firm to the manipulation of assets to be bought, sold, and re-engineered with the goal of boosting returns to shareholders. As a firm's financial success became less dependent on productive activity, managers increasingly viewed labor as another factor of production to be squeeze—rather than a reciprocal relationship that could add value to the company. The logical extension of the financial turn, as David Weil explains in *The Fissured Workplace*, is the outsourcing of larger and larger portions of the work producing goods and delivering services. Call center workers experienced the impact of the financial pressures on AT&T and Bell Atlantic to maximize short-term shareholder value over longer-term investments by growing job insecurity, downward pressures on living standards, and intensification of the stressful conditions in the call centers. In response, call center employees coalesced into a workforce of resistance, even as larger political and economic forces shifted the balance of power toward their employers (Appelbaum and Batt 2014; Batt 2018; Chandler 1977; Davis 2009; Jacoby 2021; Kalleberg 2011; Krippner 2011; Rodgers 2011; Weil 2014).

CWA and its customer service leaders and members used multiple approaches in their attempts to translate their vision of good, humane customer service jobs into concrete changes to improve the working conditions, job security, and living standards in the call centers, including collective bargaining, labor–management partnerships, regulatory interventions, public relations campaigns, and, in the year 2000, going on strike. The chapters in the dissertation explore the effectiveness of these strategies, with particular attention to the context in which they took place and the product market serviced by the call center workers. CWA made gains or defeated cost-cutting measures when the impacted customer service workforce sold higher-value products and services; when regulatory structures protected markets from downward pressure on labor costs; when their employer's financial performance was strong; when union density remained high; and when customer service members and leaders were able to mobilize the full power of the union behind their demands. In their struggles with employers, workers deployed multiple strategies in their fight for good, secure, satisfying jobs that gave them the autonomy to use their knowledge, skills, and experience to serve their customers. CWA proved to be a powerful force for collective resistance to the worst abuses that drove the unhealthy, stressful, dehumanizing conditions in the call centers and in gaining union jurisdiction over jobs of the future in new Internet and, at AT&T, wireless services. Indeed, the union made a difference in shifting the balance of power for call center workers.

And yet union power proved necessary but not sufficient to address the root causes of the insecure, unhealthy, and dehumanizing jobs in the call centers. The contest between the CWA-represented call center workers and their employers took place at a time that neoliberal ideology with its overriding faith in unfettered market forces provided few, if any, public policies to restrain the power of capital. Telecommunications policies favored nonunion new entrants. Weak labor laws were no match for aggressive anti-union employers determined to prevent their workers from organizing. Financial deregulation eliminated the few public policy guardrails on capital, leading to the financial turn in capitalism and management practices that drove cost cutting, outsourcing, intensified speedup, and stressful surveillance in the call centers. The call center workers and CWA wrestled with employers for control in an arena that was heavily stacked against them.

We have entered a new Gilded Age, one in which the 50 richest Americans hold more wealth than the combined total of the bottom half (165 million) of all Americans (Steverman and Tanzi 2020). To restore a modicum of balance for working people, we need fundamental reform of our labor laws, stronger financial regulation, strengthening of the social safety net, tax and budget policies that finance investments not only in our physical infrastructure but most important in our people, and a restoration of our democratic institutions. Working people cannot be left on their own to battle the tremendous power of unrestrained capital. Instead, we need the support of government institutions to give them a fighting chance to come together to exercise collective power, to gain their fair share of the wealth that they create, and to win a measure of control over the great power of digital technology to improve conditions in our modern automated workplaces.

As scholars, we must always reflect upon the elements of our personal background that impact the quality of our research (Reyes 2020). My study of call center workers in the legacy Bell companies draws upon my career as a member (and in the latter years, director) of the CWA Research Department (1992–2020). During this period, my responsibilities included providing assistance to the customer service occupational group in contract negotiations and enforcement, organizing, and labor–management initiatives. Over the years, I visited multiple AT&T and Bell Atlantic call centers, sat next to service representatives as they fielded calls, and was briefed by managers who demonstrated the workforce management tools they used to monitor performance. I supported the annual customer service conferences, provided research to CWA organizers at Sprint and Southwestern Bell Mobile Systems, served as an expert witness in the US West reasonable sales objectives arbitration, analyzed data for the union members of the AT&T/CWA outsourcing committee, and facilitated development of the service-representative bargaining agenda.

After passage of the Telecommunications Act of 1996, I directed CWA's telecommunications policy program with multiple interventions before the FCC, state commissions, the US Department of Justice, and Congress.

My personal background and "insider" status impacts this study in several ways. First, my female gender, my union status, and my relationship with many of the CWA call center and other union leaders opened many doors, built and sustained trust and rapport in extensive interviews, provided me access to key documents, helped me identify important struggles that informed my case studies, and facilitated introductions to call center workers and other union and management leaders who were essential to my research. For example, when I reached out to a retired AT&T sales and marketing executive, he agreed to an interview because we had worked together on the joint CWA/AT&T outsourcing committee; he then introduced me to another top AT&T executive who agreed to an interview based on his colleague's "seal of approval." These interviews helped me understand the competitive and business pressures they faced managing call center operations in the post-divestiture period. On the other hand, my union status also closed the door when another top AT&T manager declined a formal interview; it also likely played a role in the AT&T legal department's decision to refuse access to certain documents provided under confidentiality to the union.

Second, my "insider" status and background give me deeper understanding of the world of the service representative, her struggles for good jobs and meaningful work, internal union political dynamics, and the impact of regulatory change. My deep knowledge of the call center workforce and the telecommunications industry helped me ask the right questions and probe more deeply in analyzing the impact of social, economic, and political forces on the front-line workforce. My study would certainly have benefitted from access to internal company documents, managers, and executives who described the options and rationale behind key management decisions with the same level of detail as the union documents did.

Certainly, I have written "Resistance in the Digital Workplace" with a clear point of view. I believe unions empower workers, capital concedes only what labor power successfully demands, and government institutions profoundly impact the outcome of the contest between labor and capital in society and in the workplace. My goal in this project is to take advantage of this "insider" background to present a rich, nuanced portrait of the ways in which call center workers experienced the impact of neoliberalism and financialization on the job, how they fought against the degradation of conditions at work, and the constraints they faced in the context of the larger political economy.

References

Appelbaum, Eileen and Rosemary Batt. 2014. *Private Equity at Work: When Wall Street Manages Main Street.* New York: Russell Sage Foundation.

Barry, Kathleen. 2007. *Femininity in Flight: A History of Flight Attendants.* Durham: Duke University Press.

Batt, Rosemary. 2018. "The Financial Model of the Firm, the 'Future of Work.'" In *The Routledge Companion to Employment Relations*, eds. Adrian Wilkinson, Tony Dundon, Jimmy Donaghey, and Alex Colvin. New York: Routledge.

Benson, Sue Porter 1986. *Counter Cultures: Saleswomen, Managers, and Customers in American Department Stores.* Urbana: University of Illinois Press.

Boris, Eileen, and Jennifer Klein, 2012. *Caring for America: Home Health Workers in the Shadow of the Welfare State.* New York: Oxford University Press.

Braverman, Harry. 1974. *Labor and Monopoly Capital: The Degradation of Work in the Twentieth Century.* New York: Monthly Review Press.

Burroway, Michael. 1979. *Manufacturing Consent: Changes in the Labor Process Under Monopoly Capitalism.* Chicago: University of Chicago Press.

Callaghan, George, and Paul Thompson. 2001. "Edwards revisited: Technical control and call centres." *Economic and Industrial Democracy* 22: 13--37.

Chandler Jr., Alfred D. 1977. *The Visible Hand: The Managerial Revolution in American Business*. Cambridge: Harvard University Press.

Cobble, Dorothy Sue. 1991. *Dishing It Out: Waitresses and Their Union in the Twentieth Century*. Urbana: University of Illinois Press.

Cohen, Lizabeth. 2004. *A Consumers' Republic: The Politics of Mass Consumption in Postwar America*. New York: Vintage Books.

Concentrix 2020. SEC Form 10-K for the year ended December 31, 2020.

Contact Babel. No date. *US Contact Centers, 2022–2026*, 9th ed.

Davis, Gerald. 2009. *Managed by the Markets: How Finance Reshaped America*. New York: Oxford University Press.

Doellgast, Virginia. 2012. *Disintegrating Democracy at Work: Labor Unions and the Future of Good Jobs in the Service Economy*. Ithaca: Cornell University Press.

Doellgast, Virginia, and Sean O'Brady. 2021. "Collective Voice and Worker Well-Being: Union Influence on Performance Monitoring and Emotional Exhaustion in Call Centers." I*ndustrial Relations: A Journal of Economy and Society* 60, no. 3: 307–337.

Edwards, Richard. 1979. *Contested Terrain: The Transformation of the Workplace in the Twentieth Century*. New York: Basic Books.

Gabin, Nancy. 1990. *Feminism in the Labor Movement: Women and the United Auto Workers, 1935–1975*. Ithaca: Cornell University Press.

Gerstle, Gary. 2020. *The Rise and Fall of the Neoliberal Order: America and the World in the Free Market Era*. New York: Oxford University Press.

Green, Venus. 2001. *Race on the Line: Gender, Labor, and Technology in the Bell System, 1880–1980*. Durham: Duke University Press.

Hirsch, Barry T., and David MacPherson. 2005. *Union Membership, Coverage, Density and Employment from the Current Population Survey*. http://unionstats.gsu.edu

Holman, David, Rosemary Batt, and Ursula Holtgrewe. 2007. *The Global Call Center Report: International Perspectives on Management and Employment*. Ithaca: Cornell University Press.

Jacoby, Sanford. 2021. *Labor in the Era of Financial Capital: Pensions, Politics, and Corporations from Deindustrialization to Dodd–Frank*. Princeton: Princeton University Press.

Kalleberg, Arne. 2011. *Good Jobs/Bad Jobs: The Rise of Polarized and Precarious Employment Systems in the United States, 1970s to 2000s*. New York: Russell Sage Foundation.

Karasek Jr., Robert A. 1979. "Job Demands, Job Decision Latitude, and Mental Strain: Implications for Job Redesign." *Administrative Science Quarterly* 24, no. 2: 285–308.

Kessler-Harris, Alice. 2007. *Gendering Labor History*. Urbana: University of Illinois Press.

Kotz, David. 2017. *The Rise and Fall of Neoliberal Capitalism*. Cambridge: Harvard University Press.

Krippner, Greta R. 2011. *Capitalizing on Crisis: The Political Origins of the Rise of Finance*. Cambridge: Harvard University Press.

Lazonick, William, and Jang-Sup Shin. 2020. *Predatory Value Extraction: How the Looting of the Business Corporation became the U.S. Norm and How Sustainable Prosperity Can Be Restored*. New York: Oxford University Press.

Lee, Don. February 1, 2015. "The Philippines Has Become the Call Center Capital of the World" *Los Angeles Times*. https://tinyurl.com/mvvam989

Marx, Karl. 1986. "The Eighteenth Brumaire of Louis Bonaparte." In *Karl Marx: A Reader*, John Elster, ed. New York: Press Syndicate of the University of Cambridge.

Milkman, Ruth, ed. 1985. *Women, Work, and Protest*. Boston: Routledge and Kegan Paul.

Nielsen, Georgia Panter. 1982. *From Sky Girl to Flight Attendant: Women and the Making of a Union*. Ithaca: ILR Press.

Noble, David. 1986. *Forces of Production: A Social History of Industrial Automation*. New York: Oxford University Press.

Norwood, Stephen. 1990. *Labor's Flaming Youth: Telephone Operators and Worker Militancy, 1878–1923*. Urbana: University of Illinois.

Padios, Jan Maghinay. 2012. *Listening Between the Lines: Culture, Difference, and Immaterial Labor in the Philippine Call Center Industry*. Ph.D. dissertation, New York University.

Resnikoff, Jason. 2021. *Labor's End: How the Promise of Automation Degraded Work*. Urbana: University of Illinois Press.

Reyes, Victoria. 2020. "Ethnographic Toolkit: Strategic Positionality and Researchers' Visible and Invisible Tools in Field Research." *Ethnography* 21, no. 2: 220–240.

Rodgers, Daniel T. 2011. *Age of Fracture*. Cambridge: Harvard University Press.

Schmitt, John, and Jori Kandra. 2020. *Decades of Slow Wage Growth for Telecommunications Workers*. Washington DC: Economic Policy Institute, Tables 4, 8.

Sitel. May 30, 2020. http://sitel.com

Stevens, Andrew J.R. 2014. *Call Centers and the Global Division of Labor: A Political Economy of Post-Industrial Employment and Union Organizing*. New York: Routledge.

Steverman, Ben, and Alezandre Tanzi. 2020. "The Richest 50 Americans Are Worth as Much as the Poorest 165 Million Americans." *Bloomberg Wealth* (citing US Federal Reserve data).

Sykes 2020. SEC Form 10-K for the year ended December 31, 2020.

StrategyR Global Industry Analysts. 2022. https://tinyurl.com/ycx7fy9e

Taylor, Frederick. 1911. T*he Principles of Scientific Management*. New York: Harper and Brothers.

Taylor, Phillip, and Peter Bain. 1999. "An Assembly-Line in the Head: The Call Center Labor Process." *Industrial Relations Journal* 30, no. 2: 101–117.

Teleperformance. *2020 Universal Registration Document*. http://teleperformance.com

Temin, Peter, and Louis Galambos. 1987. *The Fall of the Bell System*. New York: Cambridge University Press.

TTEC Holdings 2020. SEC Form 10-K for the year ended December 31, 2020.

US Bureau of Labor Statistics. 2021a. "Employment, Hours, and Earnings." *Current Employment Survey*, December. https://tinyurl.com/y3cfy5de

US Bureau of Labor Statistics. 2021b. *Union Affiliation of Employed Wage and Salary Workers by Occupation and Industry*, Table 3. https://tinyurl.com/ybu6atrb

Weil, David. 2014. T*he Fissured Workplace: Why Work Became So Bad for So Many and What Can Be Done to Improve It*. Cambridge: Harvard University Press.

Zuboff, Shoshana. 2019. *The Age of Surveillance Capitalism*. New York: Public Affairs.

Manufacturing Apprenticeship for Dislocated Workers: A Case Study

DANIEL MARSCHALL
AFL-CIO Working for America Institute
and
George Washington Institute of Public Policy

Abstract

This case study describes the occupational background, economic context, organizational partnerships, and activities of a registered apprenticeship program held at a manufacturing firm in St. Cloud, Minnesota, during 2018 and 2019. After recounting the history of the Industrial Manufacturing Technician program, it explains the role of manufacturing in the Minnesota economy and reveals how company strategy at Electrolux Products led to the closure of a refrigerator plant and the dislocation of more than 800 workers. A table summarizes the roles of ten program partners. The registered apprenticeship program graduated 35 people in September 2019. The program demonstrated the importance of partnerships in program design and implementation and revealed how a state labor federation exercised its capacity to help establish a joint apprenticeship program. Two years after completion of the apprenticeship program, 57 percent of the graduates were employed at an average wage of $17.87 per hour, about 2 percent lower than their Electrolux earnings.

Introduction: The IMT Registered Apprenticeship

Participants in this program were enrolled in the Industrial Manufacturing Technician (IMT) apprenticeship, which is registered by the US Department of Labor as ONET Code 17-3029.09 and RAPIDS Code 2031 HY.

Initial development of the IMT was driven by the Wisconsin Regional Training Partnership (WRTP) in concert with unions, Wisconsin manufacturing employers, the state's Bureau of Apprenticeship, subject-matter experts, and a community and technical college. Founded in 1994 as a workforce intermediary (Marschall 2021), the WRTP had substantial experience working with employers to advance worker skills and maintain manufacturing employment. The organization discovered that manufacturing firms typically employed entry-level workers, generally with a high school education, who learned their production skills on the job but had limited opportunities to move up into more highly skilled, better compensated positions. The pipelines into these positions had deteriorated because of the erosion of internal labor markets and reduced reliance by companies on apprenticeship.

Moreover, as employers introduced advanced technology, the demand for high-skill workers steadily increased. WRTP staff noted that "they were hearing consistently about the need for skills inside plants throughout their network. Inside plants, the job titles were inconsistent, but the skills needed were consistent. … The IMT gives those workers a stronger base of skills, allows them to work in more skilled positions, and provides the foundation for future advancement into even more skilled positions like maintenance or electrical/mechanical positions" (Berth, Dresser, and Ubert 2017: 1–2) The IMT was seen as an innovation that would reestablish a middle rung of occupations for existing workers at manufacturers.

In 2012, the Bureau of Apprenticeship Standards (BAS) in Wisconsin coordinated occupational analysis, subject-matter expert validation, related instruction course design, and employer and union input to produce an apprenticeship program that would appeal to multiple groups, notably high school graduates, career changers, and nontraditional workers. As the BAS described these technical workers:

> Entry-level industrial manufacturing technicians operate industrial production–related equipment, work with manufacturing-related tools, and perform work processes related to a wide variety of manufacturing settings. Industrial manufacturing technician apprentices will learn to set up, operate, monitor, and control production equipment. They will also help improve manufacturing processes and schedules to meet customer requirements. (Bureau of Apprenticeship Standards 2012)

As standardized in Wisconsin, an IMT apprenticeship would stretch over 18 months—a shorter period than many other apprenticeships in factory settings—and involve both on-the-job learning and related technical instruction, often provided at community and technical colleges. Apprenticeship sponsors would be required to document 2,736 hours of training on the job and combine that with competency in a range of production processes, safety regulations, mathematics, and communication skills. The IMT was designed as a hybrid program. The necessary competencies could be documented through either 264 hours of classroom instruction or demonstrating knowledge attainment through assessment and certification as part of the Manufacturing Skills Standards Council (MSSC), which offers a validated system for identifying basic production process knowledge for front-line manufacturing workers.[1]

Advocates for the IMT apprenticeship highlight that its structure is both adaptable for employers, who benefit from a rapid increase in worker skills, and flexible in its capacity to adjust to different manufacturing settings. After agreement by employer and union representatives (when a union is present) to engage in the IMT, implementation involves consultation about what aspects of the IMT guidelines (the "Job Book") are relevant to a particular manufacturer. The 23-page Job Book specifies the duties and tasks involved in achieving ten competencies such as "Set Up, Inspect and Adjust Production Equipment" and "Measure and Inspect Work Using Mechanical Tools and Testing Equipment" (Smith 2017). The employer sponsors of an IMT program customize the curriculum to their production needs.

Once installed with several unionized manufacturers in Wisconsin, the IMT apprenticeship spread more widely through the WRTP's alliance with national organizations and access to government resources. A milestone was reached in November 2014 when the Office of Apprenticeship in the US Department of Labor approved the IMT as a registered program. Two groups worked with WRTP on expansion. One was Jobs for the Future (JFF), a Boston-based nonprofit that has created a framework for expanding work-based learning to "catalyze industry strategies" to meet skill shortages and give underserved groups greater access to credentials while supporting a family.[2] The other is the AFL-CIO Working for America Institute (WAI) (2000), which has long supported high-road, worker-centered and joint employer–union partnership approaches to work-based learning and views joint apprenticeship programs by workforce intermediary partnerships as a key strategy to increase diversity in the manufacturing workforce (Johansson 2017).

During the federal American Apprenticeship Initiative (AAI) of 2014–2015, JFF and the WAI received a federal grant and contract, respectively, to expand the IMT apprenticeship across a number of states and, later, nationally. While the two organizations set delivery strategies and administered overall program activities, state-based partnerships contacted employers and unions to implement IMT programs, provide mentorship training, deliver technical assistance, and track progress.[3] Between 2015 and 2018, these workforce intermediary partnerships registered 850 apprentices (Sullivan and Kobes 2020: 117). Electrolux in St. Cloud, Minnesota, was one of many manufacturers who agreed to sponsor the IMT. In this case, the program served dislocated workers and provided an opportunity to examine the extent to which this subgroup of unemployed workers would benefit.

Economic and Workforce Context

Manufacturing is an important segment of the Minnesota economy, accounting for about 14 percent of total output. The National Association of Manufacturers reported that there were 6,618 manufacturing firms in the state in 2016. Out of the 50 largest (in revenues) Minnesota public companies in 2018, one half of the total were manufacturing firms such as the 3M Company, General Mills, Hormel Foods, the Toro Company, and Winnebago Industries.

Although manufacturing employment declined in the decade after 2000 as a result of company outsourcing of production, it began to level off and stabilize in subsequent years (Kaul 2019). In their 2018 survey of manufacturing conditions, the state's Department of Employment and Economic Development (DEED) concluded that the outlook generally was stable. Employers were optimistic about their production for the next year, with 47 percent expecting a higher number of orders. Thirty-three percent expected their employment level to increase the following year (Olson 2019). Manufacturing employment in the state exceeded 320,000 in 2018. The Electrolux Products Freezer Division factory in St. Cloud was part of the electrical equipment, appliance, and component manufacturing sector, which accounted for between 8,200 and 8,800 workers in the state.

Manufacturing comprised the second highest number of jobs in the St. Cloud area, exceeded only by healthcare and the social assistance industry. The four-county DEED economic development region surrounding St. Cloud had an average labor force of some 238,000 people in 2018, with 25,634 jobs in 707 manufacturing firms. In turn, jobs in the region's production occupation accounted for 12 percent of total employment (Greiner 2019). Labor market analysts projected that the region's manufacturing industry would add 2,316 jobs between 2016 and 2026, with a continuing tight labor market featuring a "growing scarcity of workers" (Greiner 2019: 5).

Registered apprenticeship is a well-established model of job training in the state, where a Minnesota Apprenticeship Initiative provides up to $5,000 per apprentice to employers and is expanding the model into industries such as advanced manufacturing. There were 195 active apprenticeship sponsors statewide serving about 11,600 apprentices in December 2018 (Minnesota Department of Labor and Industry 2018–19). In partnership with community and technical colleges, the Minnesota AFL-CIO helped to establish the IMT program at manufacturers such as Hood Packaging and West Rock (with United Steelworkers Local 264) and Baldinger Bakery (with Bakery, Confectionery, Tobacco Workers, and Grain Millers' International Union Local 22).

"We've worked with the AFL-CIO and the IMT apprenticeship program to deliver [stackable credentials] in partnership," affirmed Heidi Braun of Anoka-Ramsey Community College. "To date we've worked with 40 some companies. We have quite a few number of successes with the IMT program across the state [where] about 20 companies have been working with us."

These partnerships built a foundation of knowledge and experience for the formation of the IMT apprenticeship at Electrolux Products when workers were confronted by the closure of the St. Cloud facility.

Company and Union Overview

Based in Sweden, Electrolux is the world's second largest manufacturer of consumer and commercial appliances, with some 54,000 workers in 58 countries in 2018. Founded in the early 1900s as a producer of vacuum cleaners, the company expanded through multiple firm acquisitions across Scandinavia, the United Kingdom, other European countries, and the United States after World War II. It diversified into product areas such as washing machines, refrigerators, and food service equipment. In 1997–1998, a new CEO launched an aggressive restructuring strategy, cutting costs and consolidating operations by closing 24 factories and 50 warehouses. The company dismissed some 12,000 workers (Canedy 1997). Over the next four years, Electrolux cut employment in "high-waged, unionised Europe into cheaper more deregulated (de-unionised) labour zones" in Eastern Europe, Asia, and South America (Lambert, Gillan, and Fitzgerald 2005: 269).

After merging with White Consolidated Industries in 1986, Electrolux Home Products, Inc. acquired the St. Cloud facility, making it part of their Freezer Division. It is located on a site in the Pantown neighborhood that had housed an industrial plant since 1929. About 1,600 workers there continued to produce freezers for the consumer market. Explains the union shop chairperson: "Our busy season starts when the construction season picks up, people are remodeling their homes and kitchens, and then in the fall you have the harvest and hunting seasons, so people are buying their freezers [for] a hunting shack or maybe a man cave and that's where they put all the deer and fish that they caught all season." By 1998, the factory was producing, under the Frigidaire label, 75 percent of the freezers sold in the United States.

Hourly workers at the St. Cloud facility became members of Local 623 of the International Association of Machinists and Aerospace Workers (IAMAW). Out of the eight shops in the local, the St. Cloud facility was the largest, numbering nearly 1,700 members of the collective bargaining unit in 2004, when the company's cost-cutting practices appeared. That year, Electrolux moved production of a compact freezer from St. Cloud to China. As a radio station reported:

> In 2004, Electrolux began a major restructuring, shuffling jobs from high-cost to low-cost areas, from Sweden to Hungary, from England to Poland, and from Denmark to Thailand. This summer [2013], Electrolux will begin moving jobs from a town outside Montréal to Memphis, where it will begin production of ovens in a brand-new high-tech plant. (Hsu 2013)

The Memphis plant, attracted by public incentives of more than $150 million, opened in 2014 and paid union-level wages to employees represented by IBEW Local 474. Electrolux closed it in 2019 and moved production to a $250 million new plant in Anderson, South Carolina.

Restructuring also hit Michigan and Iowa. Electrolux closed their large refrigerator production facility in Greenville, Michigan, in 2006 and moved to Juárez, Mexico, where Electrolux workers made $1.57 per hour. Some 2,700 US workers lost their jobs (Goodman 2008). In Iowa, the company shut their washer/dryer factory, idling some 2,000 workers, and moved production to the Juárez plant.

The Minnesota facility suffered a similar fate. In January 2018, Electrolux announced that it would close the St. Cloud plant and move freezer production to a highly automated factory in South Carolina. Union leaders and public officials had heard rumors about the company's plans for years. In late 2017, the mayor of St. Cloud asked an Electrolux manager whether a closing was planned, and the company delivered a "blatant response" of "no," according to notes from a closed meeting of union and public officials.[4] International union officers, both of the state's US senators, and local public officials campaigned to convince Electrolux to reconsider. Electrolux replied that they expected to close the plant around the fourth quarter of 2019.

The declared closing posed a dilemma for the company. It wished to continue freezer production in Minnesota for the next year while waiting for productive capacity at the new factory to come online. Skilled workers with employment options could be expected to start leaving after the announcement. At the same moment, said a union official: "We have a huge labor shortage in St. Cloud. We have five or ten percent more job openings than job applicants. And we have a very large skills divide." How to motivate employees to stay on the job? Or come to work at a factory that was on its way out of town?

The company and union developed several responses. The company offered a hiring bonus for "motivated production employees," a severance payment when the plant closed, and a full menu of benefits afforded by the collective bargaining agreement. The IAMAW Local 623 Bargaining Committee negotiated an "effects agreement" that included a severance package, a wage increase, continued medical benefits after layoff, and agreement to "work together to help laid off employees access government programs, transition assistance, training, and other employment" (Nolasco 2018). After the local union heard about the IMT from its international representative and the

Minnesota AFL-CIO, the company and union accepted the apprenticeship program, with Electrolux as the formal sponsor.

Although additional education and skill training could help dislocated workers obtain new jobs over the long term, and the moderately optimistic projections of central Minnesota manufacturers suggested that they would be hiring, the immediate impact of the closing on the local community would be severe. The University of Minnesota Extension and the St. Cloud State University School of Public Affairs Research Institute issued an "economic emergency report" which showed that the closing of the plant, and the dismissal of its 860 workers, would cause an estimated decline of $670 million in economic output in the three counties surrounding St. Cloud. An additional 940 jobs in suppliers would be affected, for a total loss of approximately 1,800 jobs and a loss of $103 million in labor income (Tuck et al. 2018). Small businesses also were affected negatively. A local health care provider shut down their for-profit medical clinic, for example, after losing some 4,000 patients as a result of the closing's impact on family health care coverage.

The "economic shock" would be felt acutely by residents because 585 employees (68 percent) lived in St. Cloud and adjoining cities. That report, and a survey by a state agency, would help guide a partnership of organizations to develop programs that would include registered apprenticeship.

Program Development Through Partnerships

Important facets of the program included rapid response, the coordination of state and federal government resources, and the involvement of an experienced workforce intermediary who navigated the varied guidelines of separate programs. The company did not set a specific date for the closure, so they were not required to issue a notice under the mandates of federal law.

Without a formal notice, federal dislocated worker funds from the Workforce Innovation and Opportunity Act (WIOA) could not be used immediately to provide services. In this case, the displaced workers benefited from the Workforce Development Fund, a state program that accumulates resources from a small annual assessment of employers on their taxable wages— in 2016–2017 alone, the fund collected $56.4 million (Zempel 2018). The state's Job Skills Partnership Board managed disbursements, which could go to DEED for competitive dislocated worker grants or be combined with federal dislocated worker funds.

In the wake of the closing announcement, DEED rapid response staff met with company officials to gather data on their workforce and identify the interests of workers. In cooperation with the Minnesota AFL-CIO and community service providers, DEED sponsored dislocated worker information sessions and conducted an employee survey that achieved an 85 percent rate of response. At least 78 percent of the workers were over the age of 40, the survey found, and 56 percent had worked for the company for more than 20 years. Nearly two out of three sought full-time employment. Though more than one third had earned a high school degree or GED, nearly 18 percent had not completed high school (Minnesota Department of Employment and Economic Development 2018).

"One of the things that came out in the survey is basic computer skills," explained Tammy Biery of Career Solutions. "They don't use computers on the plant floor, so it came across loud and clear. They want at least basic computer skills, which they're going to need to be able to apply for a job." The relatively low rate of education of many Electrolux workers, along with their lack of familiarity with computers, presented challenges to service delivery.

Career Solutions is a workforce intermediary housed at St. Cloud Technical and Community College. It operates the St. Cloud Workforce Center under the direction of the local workforce development board and the requirements of the WIOA. Career Solutions communicated with DEED about how to deliver services and applied to the state Job Skills Partnership Board, obtaining a two-year dislocated worker grant that provided resources for a range of services. Grant monies enabled the organization to hire nine additional staff members. Career planners advised each apprentice in the program, visiting classrooms sessions periodically and helping apprentices overcome issues (such as the availability of daycare) that could present barriers to successful completion. The grant also was essential for the apprenticeship program.

When Career Solutions and Minnesota AFL-CIO staff met with company officials, said Biery,

> We explained [the apprenticeship program] to the powers that be at Electrolux, and we thought they would say, "Well, we'll get back to you on whether or not to move forward." In the meeting, they basically said, "What's our next step?" So, they were on board quickly.

What motivated the company's positive response? "I think part of it is that in order for someone to finish the IMT, they need to stay working there until plant closure. That has got to be the biggest carrot for them to participate and support programs like the IMT," Biery continued. Union officials commented that both the company and union were excited by the IMT apprenticeship "to keep employees engaged," adding that Electrolux was "taking a beating" in the local community because of the closure, and this would help their image. In a marketing strategy meeting with several partners, the company's human resources director highlighted the apprenticeship as part of Electrolux's effort to "make sure everybody has a soft landing."[5] Biery estimated that 35 to 40 different agencies and community organizations participated in the information sessions.

Tricia Simon of St. Cloud Technical and Community College, noted that

> the level of partnership and collaboration is certainly a huge accomplishment here. If you look at all the parties involved to make this happen, that's very exciting. We always look for opportunities to help in the community, help the company, [and] help those individuals to meet their needs. It's a win–win really for everyone.

In addition to interfacing with the company, organizing information sessions for workers, and supporting Career Solutions' request for state funds, DEED submitted a petition to the US Department of Labor to qualify Electrolux workers for Trade Adjustment Assistance (TAA) For Workers services. The petition was certified in May 2018, making workers who would become "totally or partially separated from employment" eligible for benefits such as job search allowances, assistance with relocation, paid education, and extended unemployment benefits.[6] One limitation: TAA funds could not be used for staff time to coach and mentor dislocated workers or provide tutoring in basic skills while they were still on the job. Career Solutions used state dislocated worker grant funds to pay tuition for the apprentices, which the college then used to pay instructors. Table 1 (next page) summarizes the roles of major partners.

The organizations listed in the table were the most visible partners engaged in one or more aspects of the many services available to those being directly displaced. As the impact of the shutdown rippled through the regional economy, workers in other organizations were affected. The *St. Cloud Times* newspaper reported, for example, that the closure would affect the jobs of some 300 workers whose employer, a nonprofit organization for disabled persons, had a 15-year-old contract with Electrolux to manufacture parts and packaging containers for Electrolux freezers; it was not clear whether the company would obtain other contracts to retain all their employees (Ostuni 2018).

Other community-based groups emerged to provide necessary social and supportive services. Adult basic education services were required and provided by a group specializing in that area. Overall, the IMT apprenticeship was part of a complex process of identifying and mobilizing resources from a number of organizations and institutions to meet the needs of displaced workers. Flexibility and creativity were essential because those needs changed during different phases of the effort.

Table 1. Selected program partners

Organization Name	Role(s)
Minnesota AFL-CIO	Staff met directly with company and union officials to introduce the IMT and served as liaisons for the company, the union, state government agencies, and community colleges to introduce the IMT and plan for its introduction and implementation. Staff helped recruit participants and then tracked workers throughout the program, completing forms to document achievement of competencies on the shop floor.
Electrolux Home Products, Freezer Division	Formal sponsor of the IMT registered apprenticeship program. Provided time and space in a training center for IMT classes. Sponsored job fairs in September 2019 during work time to give employees access to manufacturers and community service organizations.
IAMAW Local 623	Leadership negotiated an effects agreement that included support for job training for their members. The local supported the IMT apprenticeship program and recruited members to participate.
Wisconsin Regional Training Partnership	Provided technical assistance to Minnesota AFL-CIO staff on the IMT. Participated in information sessions. The WRTP application form for the IMT was used to gather initial data on workers interested in apprenticeship.
Career Solutions	Obtained a state grant to provide comprehensive services to dislocated workers, including IMT apprentices and those eligible for TAA. Worked with technical colleges to give workers access to computers. Tracked the progress of IMT apprenticeship program. Reported provision of services to the local workforce development board and other state government bodies.
Minnesota Department of Employment and Economic Development	Sponsored rapid response sessions for workers who would be dislocated from Electrolux and recruited community participants. Conducted a survey of Electrolux workers. Prepared and submitted a petition to the US Department of Labor to certify dislocated workers as eligible for TAA services.
St. Cloud Technical and Community College	Provided office space for Career Solutions and the local American Job Center as part of the state's CareerForce network. Served as a location for workers to access computers, as needed. During the fall 2018 semester, the college employed a classroom instructor for the related instructional aspect of the IMT. Enrolled IMT apprentices as college students, enabling them to earn credits that could be used for an associate's degree.
Pine Technical and Community College	Accumulated experience working with manufacturing companies and unions to implement the IMT apprenticeship. During the 2019 semesters, employed the instructor teaching the classroom aspect of the IMT.
AFL-CIO Working for America Institute	Provided technical assistance to the union-related organizations on implementing the IMT apprenticeship. Connected the Electrolux/Machinists apprenticeship to national resources. Publicized results of the apprenticeship program. Used resources to sponsor a case study to help to determine the relevance of the IMT to dislocated workers.
Jobs for the Future	Provided work-based learning policy and strategic context for the expansion of the IMT to St. Cloud and other locations. Received a national grant to assist the formation of a registered apprenticeship in manufacturing. Reported the progress and results of the Electrolux program to the US Department of Labor.

Program Components and Activities

As outlined by the US Department of Labor, a registered apprenticeship program has five components: business involvement, related classroom instruction, structured on-the-job training, rewards to workers for the gains in skills, and the achievement of a nationally recognized occupational credential (AFL-CIO Working for America Institute and ApprenticeshipUSA 2017). The Electrolux/Machinists IMT apprenticeship included four of these components, with a modification of the first point (because the program was supported by both the business and union) and a variation in the "rewards" owing to the nature of the population and the strength of the union winning gains for all hourly workers.

As the formal sponsor, Electrolux documented the participation of each apprentice and was responsible for maintaining their records and troubleshooting any problems. The company promoted the program by discussing it with supervisors at periodic meetings. The company's human resource department staff assisted interested workers in completing and signing an apprenticeship agreement, standardized by the Minnesota Department of Labor and Industry (DLI) and available as an online form. In accord with the collective bargaining agreement, the representatives of IAMAW Local 623 were engaged in setting the terms of the apprenticeship program.

The union recruited participants while acknowledging the challenges likely to be faced by their members. The union shop chairperson and chief communicator, who had worked at the plant for some 11 years, joined the apprenticeship program and continued through graduation. When I asked about his support, he speculated that it would help motivate his fellow workers. One of the challenges for union members, he said, was that some were intimidated by the prospect of taking classes because they had not attended classroom-centered school for many years. He argued that an apprenticeship certificate, along with their years of manufacturing experience, would make them appealing to future employers.

> The moment I read about it, I was a strong supporter. I'm enrolled in the apprenticeship program. People in there keep asking me questions about it and I'm like, "I looked into this, what do you really have to lose? It's one day a week [in a class] and you're here late and you do some homework." Having this apprenticeship and having this certificate, along with decades of experience, especially your machine operators, your classified people or assembly ... this is a great leg up.

When one female worker heard about the program, she talked to her co-workers and decided it would be a good thing to do. She had worked for Electrolux for 15 years and had experience in several departments, becoming a team leader and completing the firm's management systems course. She had learned to weld while working at Crestliner Aluminum Fishing Boats and taken computer courses at a community college. Personal gratification in achievement and a commitment to continuous learning propelled her forward.

> I'm in my 50s, so I grew up with a lot of female discrimination—you can't do this, you can't do that. I couldn't even take a welding class when I was in school. They laughed at me. For me to do something like this here, I do it because I can. I like hands-on stuff. I just built a deck for my house. All by myself. That's kind of where my mind is. I keep learning. It's just part of life, and if I can get a better job than making what I make here, that's what I need to do.

After the DEED rapid response sessions had been held, a diverse planning committee established, the TAA petition certified, and the company agreeing to sponsor the apprenticeship, partners prepared an IMT recruitment and marketing plan. An August 2018 morning meeting at the company included most key partners: the Electrolux human resource department, the Minnesota DLI, the state agency DEED, Career Solutions, St. Cloud Technical and Community College, JFF, and the Minnesota AFL-CIO.[7] Reporting on the session was Filiberto Nolasco Gomez, editor of *Workday Minnesota*.[8] The human resource director announced that intake forms for the apprenticeship program were available in the HR office and that 17 people had signed up thus far. Minnesota AFL-CIO staff emphasized the sign-up deadline of mid-August. The first class would be held on September 17, 2018. Timing here was critical. A fall 2018 start of classes was barely sufficient: it would allow about 16 months for the entire program if Electrolux would leave open the plant through 2019.

That same afternoon, visiting St. Cloud from their headquarters in St. Paul, Todd Dahlstrom and Jerome Balsimo of the Minnesota AFL-CIO presented two orientation sessions. The closure was a "cruddy situation," said Balsimo, acknowledging the raw feelings of fear and betrayal shared by many workers. But the IMT apprenticeship, he continued, offered workers the chance to get a journeyworker card, certifications, and college credit that would help them get new jobs. Dahlstrom asked the group: Why was this important? He related a personal story. He had worked for 11 years in a box board factory and learned a great deal on the job about manufacturing work. When the plant closed, however, he had no verification of what he had learned. He wished that he had the opportunity to gain national skilled-occupation credentials while still earning an hourly wage. "This is one way to get an advantage" in the job search and reemployment process, he argued.[9] Also present was Tricia Simon of St. Cloud Technical and Community College, who explained that successful apprentices would obtain 16 credits at the local community college—credits that could be applied later for an associate of applied science degree—and obtain a production technologies certificate, knowledge that is a core part of the MSSC.

The most visible aspect of the apprenticeship was the related instruction held at convenient times. There were two class sessions per week, both held on Monday, organized to coincide with shifts at the factory. Importantly, classes would be held at an accessible and familiar location, a company training center on 8th Street North, directly across from the plant, enabling apprentices to attend in proximity to their shifts. Apprentices working the first shift would finish work at 3:30 p.m. and walk over to a class starting at 3:45 p.m. Second-shift workers came to the training center before work, taking a 1:15 p.m. class and going to work at 3:30 p.m. In addition to the two hours of classes, it was estimated that apprentices would spend another two hours weekly doing homework, chiefly completed on computers with the D2L Brightspace system used across Minnesota community colleges.

In the first round of applications, comprising those who expressed interest in the program, 57 people signed up.[10] The majority of those initial enrollees were Caucasian males. The group included 19 people of color and eight females. Two supervisors expressed interest. Excluding the two managers, the average wage of the others was $18.52 per hour.

To lead the classes, Career Solutions and program organizers selected an experienced college teacher who appeared to develop a good rapport with apprentices. In her summary of work experience, Mary Kay McVey emphasized her 15 years in management consulting, experience in manufacturing settings, and abilities in "effective training techniques" and "customized training" to reach "organizational goals and objectives." With an undergraduate degree in engineering technology and a master's in project management, she had taught at several state community and technical colleges, so she was well-known to decision makers.[11]

Through previous teaching and consulting, McVey had assembled an extensive collection of manufacturing-related instructional material, available in electronic formats, especially valuable in a class that would not use hard-bound textbooks. Students retrieved reading material by computer and took quizzes on the class's D2L site, which they accessed at home or at the local community college (where technical assistance was available). Much of the classwork was done in small groups, she said, to encourage peer-to-peer support. Beyond imparting necessary technical, product quality, and safety knowledge, she described how she was attuned to the special circumstances of the Electrolux factory workers. Participants in the two classes were diverse, both in their facility with the English language and familiarity with structured schooling.

> I can think of a few individuals [who have] never done school before. They were very, very hesitant of even starting this program. Their confidence was very low. They didn't think they could do this, and [I encouraged] them along the way: "You got this, you can

> do this, I know you can. These are the things that manufacturing employees do all the time. And you're taking the step forward to better yourself." ... Most of them had been there for their ... entire career, which is anywhere from 10 to 30 years plus. This is the first time [they'd] been in a [college] setting.

In several instances she modified aspects of the class and ensured that apprentices had extra tutoring to tackle difficult material, while involving them in critical decisions. Because some apprentices had difficulty completing computer-based tests, she would download the questions and go over them in class to prepare students to complete the exams online. The classes in technical mathematics were very challenging for students with English as their second language, so an adult basic education instructor held Saturday morning tutoring and peer-support sessions to review the material at an appropriate pace. When Electrolux announced that the plant would close on November 1, 2019—earlier than expected—it appeared that there might not be enough time to complete the required hours. McVey gave the apprentices the choice of when to finish, and 83 percent voted to end on September 30. This decision meant doubling up by holding two class session in some weeks, placing increased stress on incumbent workers facing an unknown employment future.

Business support for the program stretched from the Electrolux sponsorship to the participation of employers in two job fairs held in September 2019, a few weeks before apprentices graduated. Sixty organizations were there to talk with Electrolux workers, including small and large manufacturers, healthcare employers, local transportation agencies, a brewery, the local school district, a bank, and nonprofit organizations.

In a typical manufacturing apprenticeship, where skill training is tied to employee upskilling and expected advancement, apprentices receive systematic wage increases as they progress successfully through time. This practice was not applicable to this program. Rather, the union-negotiated effects agreement provided higher wages for all union members who remained at Electrolux until the closure. This aspect of the program, owing to the reality of dislocation, was distinctly different. The effect on participant motivation remains unknown.

Program Outcomes

The most immediate and tangible outcome was the receipt of completion certificates for those who had finished the classroom training and acquired necessary skills in the factory. Thirty-five former Electrolux workers participated in an apprenticeship graduation ceremony on September 30, 2019. The apprentices who obtained a journeyworker card included two women.

In the federal employment and training field, a common indicator of an effective job training program is the percentage of graduates who have obtained employment six months after program conclusion. In this case, the six-month point was March 31, 2020. Employment at manufacturing firms was affected by the coronavirus pandemic. Minnesota Governor Tim Walz issued an emergency executive order on March 25, 2020, for all residents to stay at home to limit spread of the virus. In March and April 2020, Minnesota manufacturers cut 23,600 jobs, a reduction of nearly 26 percent in manufacturing employment compared to a year earlier (O'Neil 2020). Presumably, the emergence of the pandemic, combined with the availability of TAA for Worker benefits—which enable eligible displaced workers to gain extended unemployment insurance benefits on the condition of college enrollment—influenced the career and employment choices of apprentices.

Out of the 35 apprentices who completed the program, employment and hourly wage information was available and pertinent for 33 of them (one had retired, and another had passed away from an illness contracted before the pandemic). The average wage of successful nonmanager apprentices working at Electrolux was $18.21 per hour. The six-month outcome statistics for dislocated workers were generally positive. A majority (61 percent) were reported to be enrolled in college as of March 31, 2020. Nine workers (27 percent of the 33 respondents) were employed and earning wages that averaged $17.07 per hour, a decrease of 6 percent from the previous average. About one half of the employed former apprentices were working for manufacturing firms. One worker reportedly obtained a manufacturing job and then lost it due to a coronavirus-related layoff. Five workers were unemployed and not attending college six months after the plant closing.

Though the virus continued to be "firmly entrenched" in the state, manufacturers reported that they were "busier than ever" in 2021, with 62 percent hiring for open positions and 87 percent claiming to have difficulty finding qualified workers (MPR News Staff 2021). After dropping to 301,100 in May 2020, the total number of manufacturing employees in Minnesota increased to 318,600 by October 2021 (US Bureau of Labor Statistics 2021). Manufacturing jobs apparently were available to those willing to apply for positions.

Employment data for 28 apprenticeship graduates was available on September 30, 2021, two years after the end of the program.[12] Fifty-seven percent were employed, higher than the March 2020 figure. The average wage of nonmanagerial employed graduate apprentices was $17.87 per hour, about 2 percent lower than their Electrolux earnings. Three people were employed at entry-level wages of $14.00 per hour; they could be expected to increase their earnings as they gained experience and seniority on the job. Seven employees worked for manufacturers. One was involved in healthcare services.

The proportion of displaced hourly workers in college was somewhat lower than 16 months earlier. Fourteen of the 28 successful apprentices contacted reportedly remained in college two years after program completion. Nearly all the graduates were occupied with work or schooling. Only one respondent was not employed and not taking college courses.

Why did this proportion of participants choose to stay in college? State labor federation leaders speculate that Electrolux workers gained confidence through the apprenticeship in their ability to complete college courses and be successful in a college environment. Their personal pride in going back to school may have been a motivating factor, especially considering that many were attending college for the first time. Federal TAA for Workers provided tuition coverage and other financial support for those eligible, making it easier for participants to enroll in other credentialed training programs. Additionally, a Career Solutions career planner in contact with former Electrolux workers stated that many "saw college as an opportunity to have a trade which would not only give them a career but also allow them the opportunity to support their families … [and] start at wages higher than their Electrolux wages."

Aside from quantitative measures of effectiveness, the apprenticeship program at Electrolux produced several programmatic and organizational outcomes. In their review of 44 grantee programs funded under the AAI, the National Governors Association (2020) identified strong partnerships as an important factor in successful apprenticeship programs. The Electrolux program confirmed the value of partnerships in program development and implementation. The company reacted quickly to the apprenticeship opportunity and took the important steps of sponsorship and making a convenient training center available for classes. Union support enhanced the legitimacy of the program and provided peer support for apprentices who were not accustomed to classroom education. The link with community colleges and their libraries was essential in apprentices having access to computers. A workforce intermediary organization, with a track record of serving dislocated workers and an understanding of state funding mechanisms, oversaw a varied group of services and responded flexibly as the program moved forward.

The program also demonstrated that a state labor federation has the capacity to drive a registered apprenticeship program forward though productive collaboration with partners in the context of national policy. Ordinarily, joint apprenticeship programs in an industry are established by individual labor unions working with selected employers. The Minnesota AFL-CIO developed working relationships with all the actors necessary to launch a successful apprenticeship, fulfilling a liaison function that stretches beyond any single company or labor union. State labor federations have the capacity to contribute to long-standing AFL-CIO policy and practice to boost regional economies and provide opportunities for worker advancement to increase the adoption of high-road strategies.

Methodology

Research, analysis, and writing for this case study stretched from May 2018 to December 2021. Research involved a comprehensive literature review, site visits, structured interviews, and field notes of observed program activities and events. I compiled and analyzed approximately 145 news articles, research reports, spreadsheets of participant demographics, minutes of local workforce board meetings, Electrolux corporate documents, survey results, registered apprenticeship standards, and community college curriculum and promotional materials. I conducted three site visits to St. Cloud during 2018 and 2019, which enabled me to observe classroom sessions, a marketing meeting of partners, and participant orientation. After preparing an interview schedule of questions, I conducted seven structured interviews with union officials, community college professional staff, apprentices (including a person of color and a female), and workforce intermediary staff. All interviews were recorded and transcribed. Gathering other research data involved multiple conversations with Minnesota AFL-CIO staff, extended email exchanges, and following program progress during conference calls with JFF, the AFL-CIO WAI, and workforce intermediary staff.

Acknowledgments

This paper could not have been completed without the assistance of Minnesota colleagues and advisers, especially the guidance and expertise of Todd Dahlstrom and Jerome Balsimo of the Minnesota AFL-CIO, which was headed by state labor federation President Bill McCarthy at the time of the study (Bernie Bernham was elected Minnesota AFL-CIO President in February 2022). Joe Baratta and officials of IAMAW Local 623 kept me grounded, providing their insights into why apprentices were engaged in this rigorous training, how the entire operation was proceeding, and arranging for workers to be interviewed. Tammy Biery of Career Solutions guided me through the intricacies of state and federal funding and the flow of services to participants, clearing up confusing aspects of the overall effort. Community college professionals Heidi Braun and Tricia Simon explained the way college courses, procedures, and resources intersected with educational services for dislocated workers. Mary Kay McVey of Organizational Partners allowed me to visit her Electrolux apprenticeship class and helped me understand the challenges facing workers. Electrolux company officials, particularly my main human resource department contact person, Karen Schmidt, were open and helpful, giving me access to the training center and company classroom, for example, and providing information such as the roster of employers registered for company-sponsored job fairs. The workers I interviewed were forthright in sharing their experiences and career aspirations with me. My colleagues at the AFL-CIO WAI — Brad Markell, Beth Shannon, and Jane McDonald—encouraged me throughout an extended process of research, field site visits, and writing, while Myriam Sullivan and her colleagues at JFF provided their input and communicated results to government officials. Understanding a multifaceted training and education program is truly a collective effort and I am grateful to all those involved.

Endnotes

1. Detailed production standards for MSSC certification are organized according to categories of safety, quality practices and measurement, manufacturing processes and production, maintenance awareness, and green production (see Manufacturing Skills Standards Council. Production Standards, 2017 Edition. Washington, DC: Manufacturing Skills Standards Council, 2017, https://tinyurl.com/bddp8k3t)

2. JFF regards the IMT apprenticeship as a prime example of a program that "rewards skills development and enables workers to demonstrate mastery of both educational content and competencies acquired on the job. Supervisors verify that apprentices have demonstrated mastery of competencies … allowing flexibility in completing the apprenticeship" (see Charlotte Cahill, Making Work-Based Learning Work. Boston, MA: Jobs for the Future, July 2016: 15, https://tinyurl.com/ms69ftfu)

3. The workforce intermediaries and labor federations that worked with JFF, WAI, and WRTP to expand the IMT include the Keystone Development Partnership (Pennsylvania), the Labor Institute for Training (Indiana), the Human Resources

Development Institute (Michigan), the Chicago Federation of Labor, the Ohio AFL-CIO, and the Minnesota AFL-CIO (which has formed the Minnesota Training Partnership).

4. Baratta, Joe. "Tons of Support after Electrolux Closure," February 10, 2018.
5. Field notes: May 22, 2018; June 27, 2018; August 8, 2018.
6. For those eligible workers not in the apprenticeship program, TAA funds were used to pay for college courses at several community colleges in areas such as commercial driver's license, welding, computer skills, and special career track counseling. Biery reported that more than 100 workers dislocated from Electrolux were enrolled in St. Cloud Technical and Community College programs and courses by November 2019.
7. Field notes: Marketing meeting, August 8, 2018.
8. Published by the Labor Education Service of the University of Minnesota, *Workday Minnesota* is an online newsletter that is sponsored by ten unions. Introduced in 2000, it "emphasizes long-form investigative journalism" and states that it was the first online labor news publication in the United States. The newsletter ran several stories about the union response to the Electrolux closing and the IMT apprenticeship. The newsletter is available at https://workdayminnesota.org.
9. Field notes: Two orientations, August 8, 2018.
10. The Electrolux jobs of these initial enrollees included general production (12 of the 57), assembly, electrical control technician, forklift mechanic, performance tester, silver solder operator, leadperson, EMS change agent, plastic machine operator, technician, foam platform helper, and machinist.
11. McVey is president of Organizational Partners. A description of her work experience is available on LinkedIn, https://tinyurl.com/ys84yk6j.
12. Out of the 33 counted in 2020, five could not be located.

References

AFL-CIO Working for America Institute. 2000. High Road Partnerships Report: Innovations in Building Good Jobs and Strong Communities. Washington, DC: AFL-CIO Working for America Institute. https://tinyurl.com/5ch84ku3

AFL-CIO Working for America Institute and ApprenticeshipUSA. 2017. Manufacturing and Registered Apprenticeship: A Toolkit. Washington, DC: AFL-CIO Working for America Institute, June. https://tinyurl.com/5n97a9a9

Berth, Rhandi, Laura Dresser, and Emanuel Ubert. 2017. Moving Apprenticeship into Manufacturing's Future: Industrial Manufacturing Technician. Center on Wisconsin Strategies, February: 1, 2. https://tinyurl.com/7p4env48

Bureau of Apprenticeship Standards. 2012. Industrial Manufacturing Apprenticeship: Program Design Summary. Madison, WI: Wisconsin Department of Workforce Development.

Canedy, Dana. 1997. "Electrolux to Cut 12,000 Workers and Shut Plants." *New York Times*, June 13.

Goodman, Peter S. 2008. "2 Outcomes When Foreigners Buy Factories." *New York Times,* April 7.

Greiner, Luke. 2019. Economic Development Region 7W Central. 2019 Regional Profile. St. Paul, MN: Minnesota Department of Employment and Economic Development, September 20.

Hsu, Andrea. 2013. "One Manufacturing Giant Creates Winners and Losers." National Public Radio, April 8. https://tinyurl.com/bdbvbu9w

Johansson, Erin. 2017. Advancing Equity through Workforce Intermediary Partnerships: Best Practices in Manufacturing, Service and Transportation Industries. Washington, DC: AFL-CIO Working for America Institute and Jobs with Justice Education Fund, October. https://tinyurl.com/ruy2e5f9

Kaul, Greta. 2019. "Even After St. Cloud Electrolux Closure, Minnesota's Manufacturing Economy Isn't All Bad News." *MinnPost*, November 5.

Lambert, Rob, Michael Gillan, and Scott Fitzgerald. 2005. "Electrolux in Australia: Deregulating, Industry Restructuring and the Dynamics of Bargaining." *The Journal of Industrial Relations* 47, No. 3 (September), pp. 261–275.

Marschall, Daniel. 2021. Workforce Intermediary Partnerships: Key to Success in High-Performing Labor Markets. Washington, DC: AFL-CIO Working for America Institute and the George Washington Institute of Public Policy, September. https://tinyurl.com/3w4mfj6k

Minnesota Department of Employment and Economic Development. 2018. Electrolux Rapid Response Survey Report.

Minnesota Department of Labor and Industry. 2018–19, *Apprenticeship Works*, Issue 26, Winter 2018–19.

MPR News Staff. 2021. "COVID-19 in MN: Deaths, new Cases Climb as Pandemic Stays Entrenched." Minnesota Public Radio, September 24. https://tinyurl.com/3myxn4s6

National Governors Association. 2020. Registered Apprenticeship Reimagined: Lessons Learned from the American Apprenticeship Initiative. Washington, DC: NGA Center for Best Practices, November. https://tinyurl.com/3udrvzmy

Nolasco Gomez, Filberto. 2018. "Minnesota IAM Local 623 Members Approve 'Effects Agreement' at Closing Electrolux Plant." *Workday Minnesota*, March 30. https://tinyurl.com/yckh9yj4

Olson, Magda. 2019. Stability Expected for 2019. Minnesota Manufacturing Business Conditions Survey. St. Paul, MN: Minnesota Department of Employment and Economic Development, January.

O'Neil, Sam. 2020. April Jobs Numbers Illustrate Covid-19's Impact in Minnesota. St. Paul, MN: Minnesota Chamber of Commerce Foundation's Center for Economic Research, June. https://tinyurl.com/3nv2mzh7

Ostuni, Amanda. 2019. "Impending Closure of St. Cloud Electrolux Plant to Impact Over 300 Jobs Held by Disabled Workers." *St. Cloud Times,* February 19.

Smith, Owen. 2017. Industrial Manufacturing Technician: Apprenticeship Training Guidelines. Washington, DC: Office of Apprenticeship. Office of Apprenticeship Bulletin 2015-04. https://tinyurl.com/8ps78a2d

Sullivan, Myriam, and Deborah Kobes. 2020. "Giving Manufacturing Career Pathways a LIFT." In Jerry W. Hedge and Gary W. Carter, eds. *Career Pathways: From School to Retirement.* Oxford, UK: Oxford University Press: pp. 110–124.

Tuck, Brigid, King Banaian, Richard McDonald, and Neil Linscheid. 2018. Economic Emergency Program. St. Cloud Household Refrigeration and Home Freezer Manufacturing Plant Closure. St. Paul, MN: University of Minnesota Extension, March. https://tinyurl.com/9earnmnd

US Bureau of Labor Statistics. 2021. All Employees: Manufacturing in Minnesota, October. Washington, DC: GPO. https://tinyurl.com/bdzdtw8a

Zempel, Chris. 2018. Issue Brief: The Minnesota Workforce Development Fund. St. Paul, MN: Minnesota House of Representatives, Fiscal Analysis Department, July.

V. Labor and Corporate Governance II: What Would Meaningful Stakeholderism Look Like and What Would It Take to Get It?

Managerial, Institutional, and Class-Theoretic Approaches to Shareholderism vs. Stakeholderism

RICHARD MCINTYRE
University of Rhode Island

MICHAEL HILLARD
University of Southern Maine

Abstract

We contrast managerial, institutional, and class-theoretic approaches to the debate over shareholder primacy, or what we call "shareholderism." These approaches are additive—each contributes to moving us past the pernicious prevailing view since the 1970s that the corporation has no responsibility except to its shareholders. Shareholderism has caused increasingly poor labor market outcomes since the 1980s, accelerated the climate crisis, and, according to some observers, drained enterprises of capital needed to support research and development, marketing, and product development. Managerialists argue that individual CEOs need to wake up to the "world on fire" and embrace stakeholder strategies by deploying environmental, social, and corporate governance (ESG)–informed strategies, though current evidence indicates that the S in ESG is mostly bluewashing. Institutional scholars have amply demonstrated the negative effects of shareholderism on labor market outcomes and worker power. We see the state as key to curbing the "negative excesses" through reform of corporate governance that limits predatory value extraction and empowers worker representatives. Our historical and class-centered analysis broadens the context for understanding shareholderism. We focus on class agency since the 1970s by different fractions of capital: merchant, industrial, and financial, as well as the state as both actor and mediator. Our approach—mapped across individual enterprises, national economies, and global production networks—finds that the form of financialization in the United States is exceptional; moreover, commercial and industrial enterprises in the United States draw on their own financial activities and unequal exchange with super-exploited global factory production to feed the current regime of rentiers looting domestic enterprises. We conclude by highlighting controversy within US capital over the move to stakeholderism and the potential for labor movements to take advantage of these divisions in attacking the current financial oligarchy and its deleterious effects on workers both in the United States and abroad.

Introduction

In this paper, we lay out some of the implications of three approaches to corporate governance: managerial, institutional, and class theoretic. By "approaches" we mean something close to Schumpeter's notion of vision or filter, the pre-analytic and sometimes unconscious act that sorts what it is important and what it is not important to talk about (Schumpeter 1954: 41). In the discussion of corporate governance, we see these approaches as competing but also complementary. In other words, each comes from different assumptions about how the world works, and each contributes something of value. In our view, the institutional approach has the most developed picture of enterprise governance but could benefit from a more conscious encounter with the class approach. We begin to outline that here on the specific issues of financialization and the role of shareholders and stakeholders in capitalist enterprises.

The Managerial Approach

At last year's LERA meetings, we tried to account for and evaluate the recent shift by what might be termed "big capital"—investors such as Blackrock and its chairman, Larry Fink; and the Business Roundtable (BR)—from decades of fidelity to shareholder primacy to a new embrace of stakeholder capitalism and environmental, social, and governance (ESG) investing (McIntyre and Hillard 2022). We call this version of stakeholder capitalism the "managerial approach." An academic expression of the managerial approach appears in Harvard Business School economist Rebecca Henderson's Reimagining Capitalism project and her book with that title (Henderson 2020). We find Henderson to be clear and correct in connecting shareholder primacy to environmental destruction, inequality, and political blowback against capitalism associated with a revitalized Left and populist Right. But Henderson, Fink, and the BR are almost entirely instrumental—if the right people could just be compelled to think the right way, everything would be fine. There is no sense that there are structural barriers to ESG. The political–economic vision is what we would call "neoclassical plus." In other words, there is a more or less unconscious commitment to the standard view of how the economy works—i.e., start with individual choice, consumer preferences rule, markets clear, workers are paid their marginal product, but there are institutional "frictions" that prevent optimal outcomes from being achieved. Those frictions can be removed by empowering the right people with the right ideas.

Our reading of the managerial approach is that there is a more or less sincere commitment to the E part of ESG because some capitalists recognize the existential threat that climate change poses to the long-run fortunes, or even existence, of the largest corporations. Henderson in particular is critical of the academic supporters of shareholder value essentialism, so-called agency theory, and of Milton Friedman's role in popularizing it.

Still, a close examination reveals a weak, inconsistent commitment of these ostensibly woke capitalists to improving the fortunes and conditions of their employees, a propensity to "bluewash" labor practices, and a continuing anathema for the types of regulatory reforms or independent worker organizations that would be needed to truly address labor market problems. Henderson calls for mobilizing American business as a class to unilaterally adopt high-road employment practices along with attacking climate change. Our view is that there is very little evidence of capital really worrying about or committing to improving employment conditions in the United States and that organized capital opposes any structural proposals because the anti-state, anti-union/labor reflex is in their political-ideological DNA (McIntyre and Hillard 2013).

The Institutional Approach

Over the past 30 years, a variety of scholars, both inside and outside LERA, has analyzed the collapse of the New Deal order and what might be done about it. Many of these operate within the broadly defined institutional tradition. This tradition, going back to Richard Ely, Thorstein Veblen, John R. Commons, and especially Commons students such as Edwin Witte, focuses on a number of social explanations of economic behavior, in particular the underlying rule structure embedded in institutions. Analysis of labor market and corporate behavior begins with uncovering the underlying rule structure of the relevant institutions without the presumption of some optimal solution. A good solution is the one that can be obtained through bargaining of the interested parties when those parties have equal power.

We note that very few of the contemporary practitioners in this tradition describe what they are doing in this way. Institutional economics was more of a movement than a coherent theoretical school (Rutherford 2011). As it evolved in IRRA, LERA's predecessor organization, industrial or employment relations was an eminently applied subject founded and practiced by those working with what we call the "neoclassical plus approach," such as John Dunlop, as well as truly institutional scholars such as Arthur Ross. As they came to work increasingly in interdisciplinary centers of industrial relations, following the establishment of the first center at Cornell in 1945, institutional labor economists worked alongside historians, sociologists, and scholars from an array of disciplines to solve labor problems (for more on the genealogy of the field see Hillard and McIntyre 1999). Nonetheless, we can identify an institutionalist strain in recent work on corporate governance by recent employment relations scholars.

In *The Mutual Gains Enterprise* (1994) and *Security Prosperity* (1999), Tom Kochan and Paul Osterman recognized the dramatic shift in corporate governance since the late 1970s and multiple ways in which this was bad for workers in general—unionized workers particularly—and bad even for those corporate managers embracing Kochan and Osterman's mutual gains/high-road project for dealing with the new pressures of global and domestic competition, the new regulatory environment, and new information technology. Early on, they were describing the dismantling of internal labor markets under these forces, leading to downward mobility for both union and nonunion workers. Kochan and Osterman recognized that shareholder primacy was *a*—if not *the*—driving force behind these changes.

In *Private Equity at Work* (2014), Eileen Appelbaum and Rose Batt produced a clear and detailed analysis of the effects of private equity on working conditions and living standards, including lost employment, lost pensions, lost wages, and downward mobility. Their book was also the first institutional, historical, quantitative, and qualitative analysis of the rise of private equity itself. Private equity is outside the scope of this paper, but we note that more capital is now raised privately than publicly (Rakoff 2022). We will return to private equity in future work.

David Weil's *Fissured Workplace* (2014) is an important and impactful analysis of how employment relations have changed, and it also explains the powerful implications for our understanding of corporate governance. His "fissuring" analysis, or what one of us had less concisely termed "increasing social and moral distance" (McIntyre 2008), demonstrated the specific new institutional practices by corporations that linked shareholder primacy to growing wage inequality and the debasement of employment conditions in the United States since the 1980s. Weil updated in precise terms the scope and practices we call "financialization" and revealed their connection to the attack on good employment practices. The core corporate strategy that financialized investors came to insist on is shrinking direct operations to those that support a corporation's "core competency.'" In essence, the message was, "find your distinctive niche and stick to it. Then shed everything else" (Weil 2014: 49–50).

Specifically, fissuring is the act of ending long-term employment relationships with well-paid workers, by means of a wide array of outsourcing and subcontracting practices. Pushing as much employment as possible beyond the boundaries of the firm — from the "internal" to the "external" labor market— made it possible to recruit suppliers, franchisees (such as janitorial contractors), and other supply-chain entities better able to access low-wage labor, both within the United States and abroad (Weil 2014: 87–92). In sum, the methods of core competency and fissuring became a refined method of destroying the well-paid employment of the 20th century internal labor market. This is because generous employment, in the shareholder primacy doctrine, was viewed as managers stealing profits from shareholders to be nice to workers.

William Lazonick and more recently Lenore Palladino (2021) use a theoretical framework grounded in Schumpeter and Chandler for understanding how corporate governance forms either support or undermine value creation. Lazonick's work in this area goes back to the 1980s and includes his most recent book, wonderfully titled *Predatory Value Extraction*, written with Jang-Sup Shin (2020). He has built the most detailed history and analysis of the shift from managerial to financialized corporate governance in the United States and the devastating effects of that financialization using a simple but robust formulation of "retain and reinvest" (for the managerial firm) versus "downsize and distribute" (for the financialized firm.) Lazonick and Palladino provide empirical support to LERA scholars' analysis of the devastating effects of financialization on workers. Lazonick also describes what we call a "coalition of the swindlers" that exacerbates wealth inequality at the top and centers the nefarious practice of stock buybacks and heavy dividend distribution as the main factors eating the technological/capital seed corn of the future. Lazonick and his colleagues in the Academic-Industry Research Network have worked closely with sympathetic senators and representatives in Congress to design regulatory solutions that could be one part of the answer to corporate misgovernance—for instance, through the Reward Work Act (Baldwin 2019), which would prohibit stock buybacks and give workers the right to elect one-third of the board of directors of public enterprises.

Sandy Jacoby's *Labor in the Age of Finance* (2021) focuses on how unions interacted with financial institutions and markets from the 1970s to the early 2010s, or from deindustrialization to Dodd–Frank, as the subtitle puts it. The actual terminus, in the epilogue, is the Occupy movement of 2011 that both surprised and outflanked organized labor. This period saw a resurgence in the power of financial capital and labor's attempt to harness that power to build membership and protection pensions. "The irony is that workers' pension funds supplied some of the capital that fueled financialization" (Jacoby 2021: 1), by which he means the doubling of the financial sector's share of GDP between 1970 and 2008. The shareholder value hypothesis gave unions and pension funds a new tool to replace the strike, which labor effectively lost after PATCO. Unions used their financial clout to apply pressure on top management to invest for the long term but did not see that the shareholder primacy thesis actually undermined their agenda. "The seeming embrace of shareholder interests caused some European trade unionists to be dubious of what their American counterparts were up to" (Jacoby 2021: 3). Nor did they see the connection between shareholderism and rising inequality, or that the corruption that permeated corporate America was also connected to shareholderism.

Labor's support for enhanced shareholder rights over the entire period since the 1970s has had a modest payoff at best for workers. Executives' pay soared and corporate scandals abounded. As Senator Bernie Sanders said, "The business of Wall Street is fraud" (Sanders 2015). Congress increasingly did the bidding of a corrupt and powerful financial sector. As Senator Richard Durbin said of Congress, the big banks "frankly own the place" (Greenwald 2009). Given the grim options available, one cannot really indict labor for taking its place among the advocates of shareholder value in the 1980s and 1990s. Smart people in the labor movement saw something that might be of use to them. What they lacked was an analysis of how capitalism works. Perhaps the time is come for that to be on the table as well.

The line between what we are calling managerial and institutional arguments is blurry. In general, though, the authors of the texts briefly surveyed here start by inducing the rules governing contemporary corporate behavior and use historical, qualitative, and quantitative techniques to assess the effects of these rules on working people and their organizations. As we said above, we do see value in the managerial approach but find much more here. We think a sharper focus on class could illuminate some of the lacunae in the institutional approach.

The Class Approach

Our method is based in a particular interpretation of Marxian political economy. Like institutional thought, Marxism is a fundamentally interdisciplinary approach to social analysis and comes in more than one variety. Marxists in general presume that capitalism is an exploitive system that is reproduced partly by ideology and through politics, in addition to the power of the employer in the workplace. The Marxian tradition is at least as rich and diverse as either neoclassical economics or institutional thought, and a full explication would take us well beyond the limits of this paper.

Most social scientists are at least somewhat familiar with Volume I of *Capital*, where Marx analyzes commodity fetishism, the labor process, the reserve army of labor, the growth of giant enterprise, and the origins of capitalism. In that volume, there are mostly just two actors, capitalists and workers. In the United States, as we noted earlier, capitalists have been unusually unified in their opposition to organized labor, so the Volume I analysis of capitalism works pretty well in this country in our opinion. However, to understand what has happened over the past two generations, it is helpful to extend our analysis to the rest of *capital*, Volumes II and III, where Marx presents a more refined class analysis. There Marx lays out an analysis of transportation, commercial capital, banking capital, landed property, and other groups and processes, alongside the discussion of industrial capital and workers.

We do not think anything in these volumes is the last word on the subject, nor do we believe Marx's 150-year-old analysis can be simply applied to what is going on now. But it does provide a framework that we find complimentary to the institutional work reviewed above and that adds some important caveats. There are many aspects of this, and we are developing a book-length explication of this claim, for a popular audience, including a more detailed treatment of the class model developed here as well as an analysis of the political agency of the various fractions of capital. In this paper, we examine just one aspect of the class approach—the division of the surplus under financialization.

Surplus value is the capitalist's revenue net of input and labor costs in production. The capitalist, the board of directors in a modern public corporation, is the first receiver of the surplus, but they or their designee must distribute that surplus to a variety of groups in society who provide conditions of existence for the capitalist process of exploitation in production to continue—managers, stockholders, financiers, merchants, landowners, unionized workers, the state, and so forth.

This is represented in Equation 1:

$$SV = \text{accumulation} + \text{dividends} + \text{interest} + \text{fees} + \text{stock buybacks} + \text{payments to merchants} + \text{rent} + \text{wage premia} + \text{taxes} + \text{other} \quad (1)$$

For simplicity, we reorganize this as

$$SV = \text{accumulation} + \text{financial} + \text{merchants} + \text{rent} + \text{wage premia} + \text{taxes} + \text{other} \quad (2)$$

gathering together the various payments to financiers into one group.

While capitalists tend to be unified against labor, they are split by their competition over the division of the surplus. This is particularly true in our epoch of financialization, or what Marx would call the prevalence of fictitious capital—claims to future value that have no basis in current production. The simplest forms of fictitious capital are credit money, public debt, company bonds, and company shares. On the one hand, all of these forms favor capitalist development; on the other, they can become out of control because of their relatively autonomous or fetishized existence. This dialectic is what Marx alluded to when he wrote that spokesmen for financial capital displayed a "nicely mixed character of swindler and prophet."

Capital exists only as the money actually invested in the business—and not a second time as the share of stock. But it is the value of shares that is constantly reported as if it were something real. This gets even more fantastic with the more complex forms of fictitious capital: futures, options, and complex derivatives. Fictitious capital as a whole has exploded since 1990, and government policy is more and more concerned with maintaining its value, and its values, as many observers have noted (for instance, Brenner 2020; Durand 2017; Streeck 2014; Vogl 2017).

Capital investment has declined over the long run in the major rich countries. It is tempting to see this as a kind of financiers' revenge, as dividend payments suck the resources available for productive uses. This is especially the case with the criticism of stock buybacks, a pernicious practice to be sure. But of the five biggest capitalist economies, buybacks are important only in the United States (Durand 2017: 123–125), another example perhaps, of the American capitalist class exceptionalism we have examined in past work (McIntyre and Hillard 2013). The idea that payments to finance have limited accumulation is not terribly persuasive because nonfinancial firms have increased their own financial profits more than their outflows to financiers (Durand 2017: 135–138).

It is certainly true, as Lazonick and others have argued, that new financial instruments and relationships have been developed partly to change power relationships between shareholders, management, and workers. The restructurings in the 1980s and 1990s focused on unionized businesses because that's where the money was (Baumol, Blinder, and Wolff 2005; Moody 2007). Agency theory provided the cover, claiming that firms must be run in the interests of the "owners." Of course, there is no legal theory that convincingly demonstrates that shareholders "own" the firm, nor is there any evidence that these restructurings produced anything resembling efficiency (Hillard 2020; Stout 2012). Instead, the restructurings established a new and higher profitability norm just as deeper and more liquid financial markets were making higher profitability possible. So shareholders were able to reduce their risk as it became easier to drop in and out of share ownership while workers' (and supplier firms') risk increased because of their long-term specific investment in the firm (Blair 1995; Roe 1994).

Direct connection to the financial markets give the largest financiers and commercial capitalists the ability to transmit shocks and risks to weaker players both domestic and international. Wages are eroded in subcontracting firms, ever more so the farther one gets from the center (McIntyre 2021; Perraudin et al. 2014). After what seemed like an existential crisis in 2008, financiers were back at the trough in 2010. Not so the rest of society. The shareholder value movement thus represents the offensive of big financiers and commercial capital against everyone else.

Industrial restructuring and dismantling of private sector unions have restored profits without it being necessary to increase real capital accumulation. Shareholder value has shifted attention to the short term, increasing aversion to sunk costs and financializing investment strategies of nonfinancial enterprises. Outside the particular case of the United States, where the need for stock buybacks in order to maintain returns to shareholders is a drag in some sectors, "the financial sector makes no net drain on the resources of non-financial firms" (Durand 2017: 137). It is financial profits and globalization that allow firms to distribute a larger share of profits to shareholders.

The connection between financialization and globalization is underplayed in the corporate governance literature. In our view, it is critical to explaining the capital accumulation downturn despite continued overall profitability. The system we have now resembles in some ways Hobson's and Lenin's descriptions of imperialism a century ago. A financial oligarchy is dominant in the rich countries, where workers are increasingly employed in personal services or minor industrial activities. The rich countries draw tribute from the periphery both directly in the form of profits and indirectly through unequal exchange of both industrial and agricultural goods.

The roots of this system lie in the post Bretton Woods liberalization of finance and trade and the "great doubling" of the world's labor supply in the wake of the collapse of the Soviet Union and its satellites and the market reorientation of China and India (Freeman 2005). The great doubling created a giant pool of exploitable labor. Trade liberalization made it easy to move goods from low-wage locations to the center. The elaborate system of global fictitious capital that led to so many "financial crises" was itself partly developed to shift value from the periphery to the center, value created in arms-length subcontracting arrangements between giant developed-country industrial and merchant capitalists and the thousands of new factories in East and South Asia, Eastern Europe, Latin America and the Caribbean, and Africa (Smith 2016).

Profits for a company like Walmart come from oligopolistic positions in domestic markets but also through ultimate control of international production networks (Lichtenstein 2009; McIntyre 2008). Thus, domestic capital accumulation is at least partly disconnected from domestic conditions as investment in low- and middle-income countries grows (Fiebinger 2016).

In sum then, while industrial capitalists do pay more of their profits to financiers, they more than make up for that through their own financial activities. For the biggest corporations, we can add in profits derived from international operations, including their super-exploitation of very low-wage labor directly but more often through transnational subcontracting networks, and the effects of deunionization. So, Equation 2 becomes

$$\text{SV} + \pi\text{fin} + \pi\text{global} = \text{capital accumulation} + \text{financial} + (\Delta\ \text{financial}) + \text{merchants} + \text{rent} + \text{wage premia}\ (-\Delta\ \text{wage premia}) + \text{taxes}\ (-\Delta\ \text{taxes}) + \text{other} \tag{3}$$

with financial distributions increasing, paid for by falling wage premia and falling taxes on the right-hand side in addition to the two new revenue sources on the left.

Conclusion

Since the Business Roundtable's redefinition of the purpose of the corporation in 2019, very little has actually happened (McIntyre and Hillard 2022). It would be too easy to dismiss this redefinition as merely corporate cover, given the proposals of then-presidential candidate Senator Warren, as well as Senator Baldwin and others, and the general blowback against corporations in American society that developed in the 2010s. We see the statement as more serious than that, reflecting the recognition on the part of CEOs and their academic allies that shareholder primacy is unsustainable, especially given the climate crisis.

If nothing else, the statements and activities of powerful financiers such as Larry Fink and organized capitalist groups such as the Business Roundtable have brought a serious intercapitalist conflict over corporate governance into the open. Proprietary capitalists, capitalists in the energy sector, and their political and academic allies are waging a campaign to maintain shareholder primacy (Gelles and Tabuchi 2022). It is this kind of moment of intercapitalist conflict that has been conducive to progressive change in the past. Thomas Ferguson demonstrated that a coalition of internationally oriented oil companies and banks were willing to tolerate the more moderate aspects of the New Deal if they were also able to obtain changes in US international economic policy that benefited them (Ferguson 1984). We may be at a similar point now—alas, without Roosevelt's massive majorities (until 1938) in Congress. In any case, our analysis may be useful for understanding the precise nature of these splits.

In probing for ways to take advantage of this situation we are arguing for seeing the role of the financial oligarchy in a somewhat larger and class-oriented context. It may be that the Reward Work Act, which bans stock buybacks and requires that workers have one-third of the seats on corporate boards, is the most sensible way forward in the current conjuncture. But we also know that the structure of financial power is embedded in a global system of capitalism now over 30 years old that is invisible to most people, even to some of its most active participants. Building an understanding of and social movement to oppose this system of financial oligarchy is critical. Unless and until we are able to expand state support for worker-governed enterprises through subsidies, incentives, and technical assistance and expand participatory budgeting and other forms of direct participation in public policy, corporations will find it relatively easy to roll back particular and limited reforms.

References

Appelbaum, Eileen, and Rosemary Batt. *Private Equity at Work: When Wall Street Manages Main Street.* New York: Russell Sage.

Baldwin, Tammy. 2019. "Reward Work Act." https://tinyurl.com/23sye54j

Baumol, William, Alan Blinder, and Edward Wolff. 2005. *Downsizing in America: Reality, Causes, and Consequences.* New York: Russell Sage.

Blair, Margaret. 1995. *Ownership and Control: Rethinking Corporate Governance for the Twenty-First Century.* Washington, DC: Brookings.

Brenner, Robert. 2020. "Escalating Plunder." *New Left Review* 123 (May/June).

Durand, Cedric. 2017. *Fictitious Capital: How Finance Is Appropriating Our Future.* London: Verso.

Ferguson, Thomas. 1984. "From Normalcy to New Deal: Industrial Structure, Party Competition, and American Public Policy in the Great Depression. *International Organization* 38(1): 41–94.

Fiebinger, Brett. 2016. "Rethinking the Financialisation of Non-Financial Corporations: A Reappraisal of US Empirical Data." *Review of Political Economy* 28(3): 354–379.

Freeman, Richard. 2005. "The Great Doubling: Labor in the New Global Economy." Usery Lecture in Labor Policy. https://tinyurl.com/yf63695b

Gelles, David, and Hiroki Tabuchi. 2022. "How an Organized Republican Effort Punishes Companies for Climate Action." *New York Times*, May 27, p. A1.

Greenwald, Glenn. 2009. "Top Senate Democrat: Bankers "Own" the U.S. Congress." *Salon*, April 30. https://tinyurl.com/3wbwybsk

Henderson, Rebecca. 2020. *Reimagining Capitalism in a World on Fire*. New York: Public Affairs.

Hillard, Michael. 2020. *Shredding Paper: The Rise and Fall of Maine's Mighty Paper Industry*. Ithaca: ILR Press.

Hillard, Michael, and Richard McIntyre. 1999. "The Crises of Industrial Relations as an Academic Discipline in the United States." *Historical Studies in Industrial Relations* 7: 75–98.

Jacoby, Sanford. 2021. *Labor in the Age of Finance: Pensions, Politics, and Corporations from Deindustrialization to Dodd–Frank*. Princeton: Princeton University Press.

Kochan, Thomas, and Paul Osterman. 1994. *The Mutual Gains Enterprise: Forging a Winning Partnership Among Labor, Management, and Government*. Boston: Harvard Business School Press.

Lazonick, William, and Jang-Sup Chin. 2020. *Predatory Value Extraction: How the Looting of the Business Corporation Became the US Norm and How Sustainable Prosperity Can Be Restored*. Oxford: Oxford University Press.

Lichtenstein, Nelson. 2009. *The Retail Revolution: How Wal-Mart Created a Brave New World of Business*. London: Picador.

Marx, Karl. 2004, 2006, 1992. *Capital: A Critique of Political Economy*. 3 volumes. London: Penguin.

McIntyre, Richard. 2008. *Are Worker Rights Human Rights?* Ann Arbor: University of Michigan Press.

McIntyre, Richard. 2021. "Worker Rights, Factory Inspection and Fashion." In *The Fashion Reader*, 3rd edition, edited by Linda Welters and Abbe Lillethun. London: Bloomsbury.

McIntyre, Richard, and Michael Hillard. 2013. "Capitalist Class Agency and the New Deal Order: Against the Notion of a Limited Capital–Labor Accord." *Review of Radical Political Economics* 45(): 125–148.

McIntyre, Richard, and Michael Hillard. 2022. "Brave New Corporate World? Stakeholderism Talk and the Folly of a Kinder and Gentler Corporate America." Under review at *New Labor Forum*.

Moody, Kim. 2007. *U.S. Labor in Trouble and Transition: The Failure of Reform from Above, The Promise of Renewal from Below*. London: Verso.

Palladino, Lenore. 2021. "Economic Democracy at Work: Why (and How) Workers Should be Represented on US Corporate Boards." *Journal of Law and Political Economy* 1(3): 373–396.

Perraudin, Corinne, Héloise Petit, Nadine Thèvenot, Bruno Tinel, and Julie Valentin. 2014. "Inter-firm Dependency and Employment Inequalities: Theoretical Hypotheses and Empirical Tests on French Subcontracting Relationships." *Review of Radical Political Economics* 46(2): 199–220.

Rakoff, Jed. 2022. "The Rich Get Richer." *New York Review of Books*, June 23, 64:11, 52–53.

Roe, Mark. 1994. *Strong Managers, Weak Owners: The Political Roots of American Corporate Finance*. Princeton: Princeton University Press.

Rutherford, Malcolm. 2011.*The Institutionalist Movement in American Economics, 1918–1947: Science and Social Control*. Cambridge: Cambridge University Press.

Sanders, Bernie. 2015. "After 5 Big Banks Plead Guilty to Crimes, Sanders Says Wall Street Business Model is Fraud." https://tinyurl.com/32dr4vy3

Schumpeter, Joseph. 1954. *History of Economic Analysis*. London: Allen and Unwin.

Smith, John. 2016. *Imperialism in the Twenty-First Century: Globalization, Super-Exploitation, and Capitalism's Final Crisis*. New York: Monthly Review

Stout, Lynn. 2012. *The Shareholder Value Myth: How Putting Shareholders First Harms Investors, Corporations, and the Public*. Oakland: Berrett Koehler.

Streeck, Wolfgang. 2014. "How Will Capitalism End?" *New Left Review* 87(May/June).

Vogl, Joseph. 2017. *The Ascendency of Finance*. New York: John Wiley and Sons.

Weil, David. 2014. *The Fissured Workplace: Why Work Became So Bad for So Many and What Can Be Done to Improve It*. Cambridge: Harvard University Press.

VI. LERA Best Posters

A Qualitative Study of the Nonstandard, Contingent Workforce and Their Socialization Into Client Organizations

JACQUELINE ZALEWSKI
JOHNNA CAPITANO
West Chester University of Pennsylvania

Our broad research question is How are nonstandard employees' (i.e., contractors and consultants) socialized into client organizations? To answer this, we conducted 51 structured interviews with three groups, each representing a different perspective on the question:

1. Nonstandard workers—i.e., contractors and consultants
2. Staffing-agency professionals who recruit and place nonstandard workers
3. Client organization managers who supervise nonstandard workers.

Data collection concluded in May 2022. In this poster presentation, we share information about our sample as well as preliminary analysis of one topic—the reasons people enter nonstandard work arrangements.

"I Will Be Happy, Where I Can Communicate": The Positive Role of Employee Representative Meetings on Employees' Quality at Work

EUN IL KIM
Yonse University

Employees' positive affect, work satisfaction, and workplace bullying are considered "hot potato" issues in current workplaces. To address these issues, decision makers and employees developed employee representative meetings. Still, not much research has been conducted on how employee representative meetings support the emotions and behaviors of people at work. Based on data from Korea, employee representatives meetings have positive effects on employee satisfaction. The analysis revealed that the positive affect of employees was higher compared to workplaces that had no such meetings.

VII. 2022 LERA Lifetime Achievement Awards

BERNARD E. ANDERSON
The Wharton School, University of Pennsylvania

I am deeply honored by the recognition LERA has given me in conferring this prestigious award. I cannot thank Janice Bellace enough for nominating me for the award. We were faculty colleagues and next-door office mates for a number of years at the Wharton School, University of Pennsylvania. I am a labor economist and Janice is a lawyer. Through the years, we often shared notes about our work. I was very happy to give Janice support and encouragement as she scaled the heights of a brilliant, highly productive career.

I was blessed by God on my journey from a female-headed row house in South Philadelphia to the heights of academia and public service. My widowed mother, Dorothy Gideon Anderson, inspired me to set my goals high and aspire to be the best at whatever I set my mind to do. She was my role model for hard work and the rewards hard work can produce in life.

When I was a young boy, my mother and grandmother occasionally took me and my brother with them when they went shopping at the famous 9th Street Italian Market in South Philadelphia. While there, I observed a notable difference in the amount of money white shoppers had compared with African Americans, and a difference in their shopping practices. For example, the prices were not displayed on the different products in the meat store. White customers typically told the merchant the type and quantity of meat they wanted to purchase. In contrast, Black customers typically asked the price of the product before requesting the quantity they bought. Black people were the dominant group shopping at Giordano's chicken store, where customers chose a live chicken who was decapitated and whose feathers were removed by machine. It was quite a process to behold.

The five-block shopping center had many small businesses located in three-story row buildings. The various shops sold fruit, vegetables, meat, fish, baked goods. Some businesses sold new and discounted clothing; one or two restaurants were nestled in between the shops. That was long before McDonald's and other fast food franchises came on the scene. There were no Black-owned businesses in the area.

After graduating from high school, I received a scholarship to attend Livingstone College, a liberal arts college in Salisbury, North Carolina, that was founded in 1874 by my religious denomination, the African Methodist Episcopal Zion Church. All freshmen were required to meet with the guidance counselor to select an academic field of study. I told the counselor I came to college to get a degree that would qualify me for a white-collar job, but unlike most of my classmates, I wasn't there to become a classroom teacher. She inquired about my interests, and I told her about my experience in the Italian market. She said if I was interested in business practices, I should major in economics.

I did very well academically at Livingstone College and was awarded a graduate assistantship from Michigan State University to pursue an M.A. degree in economics. While there, I met and was appointed a research assistant to Dr. Andrew F. Brimmer (Ph.D. Harvard), who was chair of the graduate program. He was not only my professor of money and banking, but he became my lifelong mentor. I was also taught by Dr. Jack Stieber, head of the industrial relations section. Dr. Stieber was a former president of IRRA. On his recommendation, I joined IRRA as a graduate student and started to attend the annual meetings in 1960.

On graduating from MSU, I took a job as an economist at the US Bureau of Labor Statistics in Washington, D.C. My first task was to document changes in union constitutions and governing practices that were adopted under the Labor–Management Reporting and Disclosures Act of 1959, also known as the

Landrum–Griffin Act. I also wrote and published an article on Black employment in the federal government, my first article published in a refereed journal. I was drafted and spent two years in the US Army assigned to the Department of Defense Damage Assessment Center in the Pentagon. I returned to BLS when I was discharged.

On the advice of Dr. Brimmer, who had been appointed to the Federal Reserve Board of Governors by President Lyndon B. Johnson, I applied for admission to the Ph.D. program in economics at the University of Pennsylvania. Based on my academic record, I was awarded a fellowship that paid tuition, fees, and a stipend. In three years, I completed all coursework, passed the preliminary exams, and wrote a dissertation titled "Negro Employment in Public Utilities: Electric Power, Gas, and Telephone Companies" (1969). The study built on Gunnar Myrdal's classic study, *The American Dilemma* (1946).

Dr. George W. Taylor—the endowed chair and professor of industrial relations, father of the grievance arbitration system, former chair of the Wage Stabilization Board during the Korean War, and a co-founder of IRRA—was one of my professors. I did well in both of his seminars. Several weeks before commencement, Dr. Taylor asked me to meet with him. During the meeting, he asked if I would accept an appointment as assistant professor on the tenure track at the Wharton School. I was stunned by his offer; I did not aspire for an academic career. My goal in earning a Ph.D. was to return to Washington, get a job as an economist, and work my way to the top of a civil service career. After consulting with Dr. Brimmer about the academic life of an African American in an elite Ivy League institution, I agreed to accept Dr. Taylor's offer.

I received a three-year appointment that was renewed for a second term. During that time, I taught labor–management relations and labor economics, wrote two books, and published four articles in refereed journals and several chapters in edited books by other authors. In 1976, I became the first African American awarded tenure in the Wharton School.

Throughout my career, I studied racial inequality in American economic life, seeking to understand and explain the broad, deep, persistent racial disparities in employment, income, and wealth. In addition to my academic career, I sought opportunities to put my ideas to work in proposing ways to eliminate the disparities. I served five years as a program director of the Rockefeller Foundation and six years as Assistant Secretary of Labor for Employment Standards Administration, where I had responsibility for the Wage and Hour Division, the Office of Contract Compliance Programs (affirmative-action enforcement), Office of Labor–Management Cooperation, and Office of Workers Compensation. The Labor Department job was the most difficult and time consuming I ever had, but it was a joy working as a subcabinet officer in the Clinton administration. In 2001, I returned to the Wharton School in an endowed chair, the Whitney M. Young, Jr. Professor of Management.

Through my research and observation, I concluded that African Americans have a relationship with the American economy similar to the caboose on the train. When the train speeds up, the caboose speeds up. When the train slows down, the caboose does likewise. But in the natural order, the caboose never catches up to the engine.

The unemployment rate of Black workers has been at or near twice the unemployment rate of white workers since the statistics were first collected by race in 1954. Black workers have long been the last hired and first fired over the course of the business cycle. Among the factors responsible for that pattern of turnover are seniority policies widely followed in both organized and nonorganized workplaces.

The median income of Black households has never been more than 60 percent that of white households. The ratio grew slowly following the massive migration of Black workers from the South to other regions of the United States. But the racial income disparity has not narrowed measurably in the past three decades.

The widest economic disparity is in wealth—net worth. According to data in the Federal Reserve Board income set, the median net worth of white households is $113,000 compared with less than $25,000 among Black households. Nearly one third of Black households have negative net worth—i.e., their liabilities exceed their assets. Fewer than 10 percent of white households are similarly situated. The size of the Black middle-income cohort has steadily grown over the past four decades, but wealth disparities have not narrowed at a rate commensurate with the increased number of families in the middle-income sector. It is increasingly clear that racial income, employment, and wealth disparities exacerbate racial tension and political discord. Moreover, the disparities constrain macroeconomic growth. The income disparity limits Black consumer spending, saving, and investment—major factors that drive economic growth. The US economy would be more stable if racial economic disparities are reduced, and economic opportunity is made more equal. That is a goal I will continue to support and advocate. The LERA Lifetime Achievement Award encourages me to stay the course. Thank you.

Bernard E. Anderson is the Whitney M. Young, Jr. Professor Emeritus at the Wharton School of the University of Pennsylvania, where he was the first African American tenured professor. He was Assistant Secretary of Labor during the Clinton administration and is a member of the Board of Trustees of Tuskegee University.

VII. 2022 LERA Lifetime Achievement Awards

LISA M. LYNCH
Brandeis University

I want to thank everyone, especially David Weil, for their very kind words and the honor of receiving this recognition of my work. I am especially thrilled to receive this honor alongside Professor Bernard Anderson. I was fortunate to have worked with Bernie when we both served in the US Department of Labor in the 1990s. Professor Anderson showed how an academic could leave the ivory tower and put research into practice when he served as the assistant secretary of employment standards at DOL. He was an inspiration for me as we worked together in the Clinton administration to develop and implement policies that generated more and better jobs in our economy, especially for young Black workers. Professor Anderson and I also shared the privilege of working closely with the late Professor Phyllis Wallace, a professor of industrial relations at MIT's Sloan School of Management, a former IRRA president, and the first woman full professor at Sloan. She broke through dual discrimination barriers as an African American and a woman and contributed path-breaking work as a scholar and policy maker to open up the labor market to women and minorities. But what I think she enjoyed most was mentoring the next generation of scholars and practitioners.

I was one such beneficiary of her mentoring, and she supported my development as a labor economist from my early days as a student at Wellesley college to later on when I joined the faculty in industrial relations at MIT.

In accepting this award, I want to recognize the many ways in which LERA and its previous incarnation—the Industrial Relations Research Association, IRRA—have been a critical resource for me throughout my career as a scholar, teacher, policy maker, and senior leader in higher education. Early on in my academic career, I was able to publish my work in the IRRA proceedings on racial and gender differences, when more traditional journals were reluctant to publish such research on what was viewed at the time as less important workers. Participating in the national and local chapter meetings helped me build a broad network of colleagues whose research and willingness to share real life labor–workplace experiences benefitted my scholarship and policy work. The many LERA panels and forums on collective bargaining helped inform how I engaged as provost at Brandeis University with our part-time faculty and graduate students as they first voted to unionize and then as we worked together to develop our first collective bargaining agreement.

The need for LERA has never been greater than now if we want to secure a more shared prosperity in this country. Given the current political discourse and the ease with which we can isolate ourselves in our own echo chambers, we need to find new ways to pull together rather than pulling apart. Rising inequality and increased concentration of wealth run the risk of corrupting our democracy.

This is where an organization like LERA can and must play an important role. As the only organization in the country where labor, management, government, academics, advocates, and neutrals meet to discuss and learn about all aspects of labor and employment relations, we are an important venue for the development of new ideas and frameworks to advance a more inclusive economy. We all need to continue to support and expand the convening role that LERA can play at the both the local chapter level and nationally with our social partners. As I look at the depth and diversity of our program for this meeting I am encouraged and inspired by all my LERA colleagues and optimistic for our future.

Again I thank you all for this very special honor. I want to especially thank Wilma Liebman for her exceptional leadership of LERA and making sure that COVID did not thwart the important ways we listen and learn from each other.

Discussant Comments on *Gender Segregation and Earnings in the Federal Workplace*

Doctoral Dissertation by
TERESA KROEGER
Ph.D. Candidate, Brandeis University

Discussant
VALENTINE FAYS
Ph.D. Candidate, University of Mons – Université Libre de Bruxelles

Global Overview

Kroeger's study (based on her dissertation, "Gender Pay Differentials in the Federal Civilian Workforce from 1973 to 1997") looks into the gender pay gap in the public sector in the United States using a unique federal and longitudinal dataset over a 25-year time span (1973–1997). The study analyzes whether differences in measurable worker and job characteristics explain differences in pay between men and women in the federal workforce in particular. The analyses identify whether changes in these characteristics over the period reduce women's pay differential with men by looking at the gap at the beginning and the end of the dataset.

The author conducts a literature review about previous empirical work studying the gender wage gap in the federal workforce of the United States corresponding with the period studied in her paper. Then she presents descriptive characteristics of her dataset that allow her to follow all federal workers through time. She then regresses the variable gender on the log of basic pay, controlling for numerous control variables and using time fixed effect. She finally refines her analyses by testing the impact of worker origin and of pay plans on the gender wage gap.

The findings reveal gender earnings inequity in the federal workforce across the studied time span. The gender wage gap was found to be decreasing between 1973 and 1997, to be dependent of worker origin and of the pay plan under which workers are paid.

The main value added is Kroeger's dataset, which enables her to follow workers through time and to use year fixed effect to study the gender wage gap in a work environment that is not that often studied in terms of wage gaps. This allows Kroeger to more significatively study the gender wage gap in the federal American workforce than previous work on the matter—i.e., to follow workers over 25 years. Another value added of her work is the use of moderating variables to study the gender wage gap more thoroughly according to workers origin and to the federal plan under which workers are paid.

Major Comments

In her literature review, Kroeger mentions several publications that address the gender wage gap in the federal workforce of the United States corresponding with the period that she analyzes. It should be interesting to see whether any other empirical work studied the same topic later for the United States (other than the ones already mentioned) and also if any work had been done on other countries and whether those studies found similar or contradictory results. Because the author analyzed the situation for the United States, perhaps she should justify the reasoning behind that choice. Why is it interesting the depict the situation in the United States and not in another country?

Also in the literature review, it could be helpful to bring more economical theoretical or empirical intuition about why it is interesting to study gender wage gap interacted with race and with pay plans.

In the methodology section, Kroeger states that she uses the log of the nominal basic pay of workers as the dependent variable. She sets aside supplements, adjustments, overtime, and other additional payments. Is this because the database did not provide such information? Or was it a deliberate choice? If so, why? Indeed, using gross wage in place of/in comparison with basic pay could show that, if we consider all pay components of wage, the gender wage gap in the federal workforce is even larger than the one obtained when using basic pay. For example, we could expect women not to work overtime because of family obligations, leading to an even larger wage gap between men and women in the federal workforce.

Regarding all equations and linked results presented in the paper, it is not clear whether the author used a fixed effect estimator or time fixed effect in an ordinary least squares estimator. In any case, it would be beneficial to show fixed effect results.

As to the equations and models used to decompose the gender wage gap, it could be useful to try the Oaxaca–Blinder (1973, 1973) decomposition method. Indeed, that method decomposes the wage gap between an explained and an unexplained part. In both parts, some of the wage gap is linked to explanatory and control variables used in the model. This means that even the unexplained part of the wage gap is linked to the explanatory and control variables used. Another advantage of the Oaxaca–Blinder decomposition is that the unexplained part is often related to wage discrimination.

Regarding control variables I have several questions/comments. First, in order to estimate the gender wage gap in general in Section 5 of the paper, I believe it would be better if the author directly inserted information about race and pay plan in the control variables. This would lead to adding new Models 11 and 12 to the section. Indeed, if she wants to estimate the gender wage gap, it could be interesting to control for race (that could also be a proxy of language and then help refine the estimation) and for pay plan. Then, when estimating the role of race and of pay plan on the gender wage gap in Sections 6 and 7, it would be better to show only results where all control variables are used (including race and pay plan) and then directly analyze interaction terms with gender according to the session. I think it would give a clearer benchmark picture of the gender wage gap with Model s11 and 12 previously mentioned and a provide a simpler analysis of Sections 6 and 7.

In Section 7, the author inserts a variable called "payplan equals 1 for General Schedule and Equivalently Graded (GSEG) plans (omitted), 2 for prevailing rate plans, 3 for other white-collar plans, and 4 for unspecified plans" in her model. However, in Table 7, it seems that she uses dummy variables for each category of plan, omitting the GSEG one.

For the results, in Section 6, Kroeger says that "the largest adjusted pay gaps in comparison to white men are found among Black women and Native women (both –0.11)." Are these women paid less because they are women or because of their race? Or could it be both? It would be interesting to look into that question and perhaps to dive into the literature about the double wage penalty for women from foreign origins in the labor market.

Finally, for all equations, I wonder about a possible endogeneity bias that should be addressed. Indeed, wages and being a woman can be inversely correlated, or correlated to unobserved characteristics that are not included in the model.

Minor Comments

First, in the value-added section, the author mentions that her study updates the literature for the 21st century. However, the dataset covers the 1973–1997 time span. I would suggest clarifying what she means.

Second, coming to the methodology section, the author states that she uses the log of the nominal basic pay of workers as a dependent variable. Would it be more accurate to use the real basic pay instead of the nominal basic pay?

Then, regarding Figure 1, it would possibly give greater information about wages evolution to the reader if the wage distributions were shown between 1973 and 1997.

Also, in Section 5, Model 8, when adding a variable about county and state, it would be interesting to mention which ones pay women less and which ones pay them more.

Finally, in Section 7, the author shows interaction terms for Model 21 only, with no other control variables. It might be more interesting to show interaction terms for Model 27, where all other control variables are included in the equation.

References

Blinder, Alan S. 1973. "Wage Discrimination: Reduced Form and Structural Estimates." *Journal of Human Resources*, 8(4): 436–455.

Oaxaca, Ronald L. 1973. "Male–Female Wage Differentials in Urban Labor Markets." *International Economic Review*, 14(3): 693–670.

VIII. LERA Best Papers I: Gender and Pay

Discussant Comments on *A Pay Scale of Their Own: Gender Differences in Variable Pay*

Paper by
JASON SOCKIN
University of Pennsylvania

MICHAEL SOCKIN
University of Texas at Austin

Discussant
GUILLAUME VERMEYLEN
Warocqué School of Business and Economics

Global Overview

This paper focuses on the pay gap between women and men—and more specifically on variable pay differences. The authors highlight a persistent gender gap in variable pay that is ever present across occupations, industries, and countries. This gap cannot be explained by differences in worker observables, latent ability, or income growth.

As Sorkin and Sorkin wisely mention in their manuscript:

> The increasing of variable pay in the [United States] and the lack of transparency about employee bonuses [make] investigations on whether and to what extent women may suffer from inequality necessary and crucial.

This is clearly an under-explored topic, mainly due to lack of data.

We found this paper to be interesting, well documented, and well written. It is clearly positioned in the existing literature. The authors went far in their analyses by adding new insights on the explanations for earning gaps. The policy implications are straight and nicely developed. They clearly deserve that space.

Major Comments

Our first concern is related to all the comparisons between coefficients the authors are making: Are these coefficients statistically different? In other words, we wonder whether they ran tests of differences between means. This is crucial because the paper is based primarily on comparisons of coefficients between the models.

A second major comment is related to the methodology of the paper. We think a clearer methodological section may be needed. It is not clear enough which regression method the authors use until the presentation of the tables when we find out that they ran OLS and fixed-effects models. We think this should be mentioned in the methodological section of the paper.

A third comment is related to the previous one. Fixed-effects models may still lead to some biases. For example, such models allow the authors to control for time-invariant characteristics that may influence the relationships, but they do not allow them to control for reverse causality or simultaneity. Maybe some robustness tests could be implemented—or at least a few words about this potential issue would be useful.

The next comment lies between major and minor ones and is related to another potential variable that may explain much about the investigated gaps; that is, the mismatch situations of workers. Mismatch is when a worker gets a job that does not correspond to his or her level of education and/or training (most often, it is overeducation for the job). It has been shown in the literature that educational and/or skill mismatches may explain differences in earnings. Also, women are more likely to suffer from this phenomenon.

Put together, overeducation may be one crucial moderating variable explaining why women are face discrimination in pay to a larger extent than men do. We raise that point so that perhaps the authors would be aware of that.

Minor Comments

The first minor comment is related to the data. The authors mentioned that earnings of "less than 200" are deleted from the data. Do they mean less than $200? Also, why this level as a cutting-off point?

Also, the authors excluded the top 0.01% of earners from the analysis. Why this level again? Is it trivial in this literature?

Finally, one way to increase the R^2 of the regressions is to include the lag of earnings outcomes as an independent variable. Maybe it could be tested because such a lagged dependent variable may capture lots of information.

Discussant Comments on *How Economic Development Tax Incentives Constrain K–12 Education*

Paper by
CHRISTINE WEN
Texas A&M University

Discussants
GUILLAUME VERMEYLEN
LOREDANA CULTRERA
Warocqué School of Business and Economics

Global Overview

This paper deals with tax abatements and their impact on school revenues. Globally, it shows that tax abatements could undermine educational spending and suggests limiting the power of local governments in abating school revenues.

This paper is really interesting, robust, well documented, and well written.

We develop our comments by dividing them into the major ones we think the author should consider, and some additional minor comments.

Major Comments

Our first concern is related to the methodology, which is one of the main value-added aspects of the paper. The author mixes both quantitative and qualitative analysis. However, the qualitative analysis doesn't appear until the end of the paper. We suggest creating a global methodology in two parts: the quantitative regressions and the qualitative analysis (through a focus group).

A better explanation is needed about what happens during the focus groups, what are the methods, etc.? This is a part of the paper that deserves lot more space.

A second major comment is also related to the methodology of the paper. As noted in the methodology section, if the author excluded 15 outliers from her analysis, why do they appear in the analysis and results? We propose two options here: either delete the discussion of outliers from Section 4 (Results) or revise the discussion about outliers to avoid stating that they were excluded from the analysis.

As third major comment: We think one theoretical point is missing—that is, the fact that poorer schools means lower salaries. Even if it could be seen as trivial, this should be more clearly explained in the literature review.

Minor Comments

The first minor comment is related to the regression: The author relies on OLS. But did she conduct any robustness tests? The results could be biased as a result of unobserved time invariant variables. Perhaps the author should run some of them, or at least mention this drawback. Also, does the author have any information about the quality of the adjustments through, for example r-squared, data?

A final point: The author mentions in that “distressed localities are more likely to offer incentives because they are desperate to attract businesses.” Could we imagine the reverse ? Tax abatements are mainly an attractive tool of the richest localities because they can afford them.

IX. LERA Best Papers II: Job Quality

Decent Employment and Poverty Alleviation for Socioeconomic Development and Its Implications for the Well-Being of the Citizenry in South Africa

OLANIYI "JOSHUA" OLABIYI
University of the Western Cape

Abstract

The need for governments and private employers to adequately provide decent work within the economy for all its inhabitants cannot be over-emphasized. This imperative is even more important since most obtainable work has been characterized by many detrimental dimensions that can be considered as constituting "indecent employment." From the viewpoint of human development, the paper examines how "decent employment" can serve as an antidote to poverty. Thus, decent employment can positively affect both material and nonmaterial social development, which includes health, education, social security, food security, and overall well-being. The present paper is borne out of the desire to empower the average South African citizen in specifically attaining an improved socioeconomic living standard. This study employed a qualitative, thematic analysis of selected reported cases of perceived "indecent" or nonmeaningful employment from both informal and formal sectors. Additionally, this paper highlights instances in which employees have experienced challenges in getting "dignified" or decent employment as a result of casualization, outsourcing, short-term contracts, and temporary employment. This interpretive, qualitative approach was adopted to put forward somewhat empirical evidence of the potential beneficial effect of decent employment on human and socioeconomic development. The main contribution of this paper is that it foregrounds the need for decent employment of the workforce in addressing the three-pronged societal challenges of unemployment, inequality, and poverty. The paper posits that decent employment significantly contributes to national socioeconomic development and poverty alleviation or eradication.

Introduction

"Indecent" employment practices lead to inadequacy in the sense that a seemingly employed individual is unable to meet the minimum needs for a minimum standard of living. In order to fulfill societal obligations in areas such as health, education, social security, and food security, productive employment is always a means of achieving material as well as nonmaterial social development. In this paper, and consistent with the conceptualization of the International Labour Organisation (ILO 1999), Deranty and MacMillan (2012), and United Nations Sustainable Development Goal 8 (UN SDG- 8)(sdgs.un.org/goals/goal8), "decent employment" or "decent work" is seen as "the desire of people in the world of work (whether formal or informal) to be adequately financially, materially, or otherwise compensated sufficiently to cater [to] their personal well-being and also to be able to make positive contributions to society therefrom." Furthermore, in this paper, the phrases "decent employment" and "decent work" are used interchangeably to refer to the same concept.

Meaningful or decent work has long been widely acknowledged as one of the main avenues available to people through which they can escape poverty. Thus, as Narayan et al. (2000) averred, decent employment should provide an adequate and secure standard of living among its other monetary and psychosocial benefits. They go on to argue that productive work that helps workers to avert poverty is a universal desire of workers (Narayan et al. 2000). While it would appear in some instances that people are employed, a closer examination would reveal that a significant proportion of such employments are casual, outsourced, short-term, and/or temporary. This situation is compounded when one adds the nature of employment in the informal sector, where cases of inhumane treatment and employment insecurity abound (Bell and Newitt 2010). Thus, it can be argued that poverty alleviation through employment goes beyond the mere provision of work opportunities to include providing decent employment that addresses the negative elements and factors associated with work. It is this holistic approach, it is believed, that would lead to socioeconomic development and alleviate or eradicate poverty.

The South African employment environment has not been spared the challenges highlighted above in spite of the "democratization" of the political system since 1994. Some of the challenges can be traced back to the residual effects of the prior, pre-democratic socioeconomic dispensation that had institutionalized racial inequalities and disparities in employment. To address these, the post-apartheid government has attempted to enact various labor or employment legislation. The impact of this legislation on employment and socioeconomic development has varied. A visible shortcoming of these efforts is reflected by the high rate of unemployment among youth, increasing casualization and outsourcing of employment. One argument that has been advanced for this continued dire state of affairs in employment in South Africa is that the levers of control of the employment landscape in South Africa are yet to be democratized. A corollary to this argument is the need for both the government and private sector employers to provide meaningful, honorable, and well-paying jobs that are able to meet the basic socioeconomic needs of workers. Jobs that are not able to meet these needs end up having the unintended effects of engendering negative health and nutritional outcomes and of further impoverishing the population. The increasing general population poverty rate in the country has been correlated to wage inequality and the burgeoning rate of dependency of the population on government social welfare schemes (Isaacs 2016).

Venter and Levy (2014) identified high crime rates, social unrest, and racial and gender disparities as additional factors that contribute to the malaise of indecent work in South Africa. It has been documented that these disparities negatively affect working people and their families, as well as the national economy. The absence of a national minimum wage policy has not helped the situation either, even though the Basic Conditions of Employment Act of 1997 allowed some vulnerable sectors to set minimum wage. Recognizing that the minimum wage alone cannot solve the nation's wage inequality or poverty problems, the national minimum wage level must be cognizant of the living wage necessary for a dignified life (Isaacs 2016). Based on the low-earning profiles of the majority of South African families, it is clear that current wage levels do not meet the needs of families and are not sufficient to raise families out of poverty. In the bottom half of households with a wage earner, wage income constitutes the primary income source (Isaacs 2016).

The efforts by the government since 1994 to address these issues from a macroeconomic policy standpoint have included implementation of the National Development Plan and establishment of the Decent Work Commission (Venter and Levy 2014). Unfortunately, in spite of these initiatives, the economy of South Africa has continued to experience low-level growth and a persistent increase in unemployment as a result of numerous rigidities of economic and employment reforms (Johannes 2012). Having productive employment is essential for achieving fair globalization and reducing poverty. The ILO has developed a "community of work" agenda that centers on job creation, worker rights, social protection, and social dialogue, with gender equality as one of the cross-cutting objectives. It has become a universal human right to be able to have decent work, and the principle has been incorporated into major human rights declarations.

As stated earlier, while many efforts have been made, little has been accomplished because of the high levels of unemployment. The rise in joblessness in South Africa has not been addressed. This situation somewhat negates any efforts toward ensuring decent employment because people have to be gainfully employed before concerns about decent employment can have relevance. Thus, for example, the scrutinization of the fairness, or otherwise, of a contract of employment or an analysis of the quality of the work experience can only make sense if individuals have a job to begin with.

With the above in mind, therefore, this paper attempts to explore how decent employment can be examined by first addressing the concerns about widespread poverty, disempowerment, and racial inequalities that pervade South African society. The questions to be answered are the following:

- In what way can decent work contribute to alleviating or eradicating poverty in South Africa?
- Has historical racial inequality negatively affected the possibility of attaining decent employment for majority of South Africans?
- What is the likely impact of decent employment on the socioeconomic development of South Africa?
- What implications will all these initiatives considered together have for the well-being of South Africans?

Aim

The aim of this study was to examine whether decent work can contribute to alleviating or eradicating poverty and thereby improve the socioeconomic development of society and the well-being of individuals. Along with this aim, an attempt was made to look at how legislative efforts pertaining to decent wages, safe working conditions, and the protection of labor rights have fared in shaping decent work in the country.

We hope to determine whether decent work can eradicate poverty among South Africans by providing a sustainable income for an average resident and their family. To achieve this goal, however, financial investment in citizens, particularly the most vulnerable ones in South Africa, should be a priority to improve their standard of living. Studying this issue relates to improving the effectiveness of government policy to deliver decent work that includes decent wages and salaries as well as ensuring safe working conditions, providing job security, and protecting labor rights. As a result of this study, it is hoped that it will point to ways to produce fruitful work that promotes good health, food security, affordable education, and human dignity. Programs and public policy interventions that will increase decent employment in South Africa should continue to be made available to everyone within the country.

Brief Literature Review: "Decent Employment"

As stated in Article 23, Section 1, of the Universal Declaration of Human Rights, the United Nations' approach to full employment and decent work calls for every individual to have a good job, to freely choose employment, to enjoy equitable and just working conditions, and to be protected from unemployment. The definition of "decent employment" recognizes that employment is a catalyst for more than just income but also for dignity, family stability, peace within the community, and economic growth that allows for more productive jobs and employment (Cohen and Moodley 2012). According to Wilson (1996), obtaining stable and secure work is related to physical and mental well-being as well as greater cohesion in communities (Paul and Moser 2009; Swanson 2012). The aspirations of people are best summed up by decent work. This includes productive and equitable work, a safe working environment and social safety nets for families, opportunities for people to achieve their potential and participate in society, freedom of speech, organization and participation in the decisions affecting their life, and equity for women and men.

As identified by the ILO, there are basic principles of decent work that need to be adhered to, such as enhancing the standards and rights of employees in the workplace; protecting rights, dignity, equality, and fair labor practices in the workplace; enhancement of job quality through the creation of new jobs and the improvement of pay prospects; the eradication of poverty and inequality demands; the creation and expansion of collective security and social safeguards; and promoting public discussion and negotiation among employees, employers, and the government (ILO 2021). These principles establish a connection between decent employment and poverty alleviation or eradication in societies. In support of these principles, the South African government has pledged its commitment to the attainment of decent work and sustainable livelihoods for all workers and has undertaken to include decent work imperatives in national development strategies. It is imperative for policy makers to find solutions that help improve the living conditions of their citizens. It is vital to develop strategies to improve the socioeconomic environment as part of the campaign for decent work.

Employment and Unemployment Patterns in South Africa in the Past Decade

Historically, the informal and formal employment sectors have both dominated the South African labor market. Informal sector employment involves individuals that engage in largely unregulated, and sometimes hazardous, employment, while the formal sector is often the standardized and regulated form of employment. The informal sector has been growing compared to the formal sector, which has experienced some stagnation as a result of existing labor market policy within South Africa's employment environment (Altman 2002; Edwards 2001; Meth 2001).

A summary of the South Africa informal and formal sectors is provided in Figures 1 and 2. According to the report, there were job losses in five of the eight formal sector industries in South Africa. There was a decline in employment in some aspects of the formal sector as follows: the financial sector (229 000 jobs), community and social services (151 000 jobs), manufacturing (105, 000 jobs), trade (42,000 jobs), mining (2,000 jobs), and construction (112,000 jobs). Curiously, the transportation sector (316,000 jobs) recorded the highest job gains, while utilities saw no change in employment.

Figure 1. Formal sector employment in South Africa

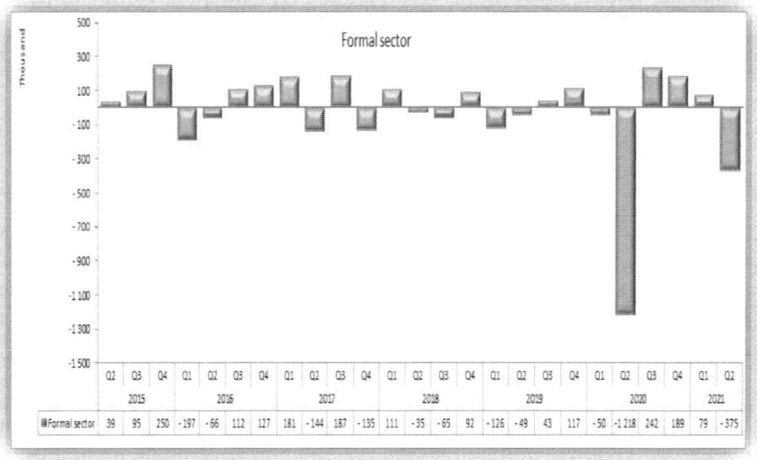

Source: Quarterly Labour Force Survey, Quarter 2, 2021.

Figure 2. Informal sector employment in South Africa

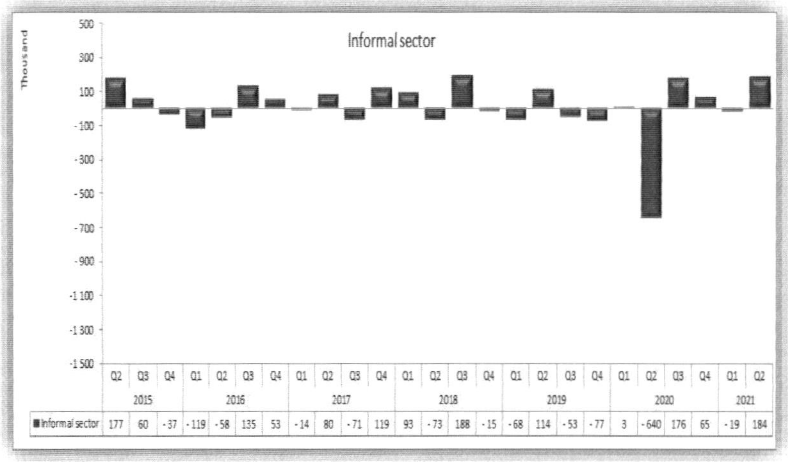

Source: Quarterly Labour Force Survey, Quarter 2, 2021.

The representation of the South African formal and informal sector employment patterns in the above figures indicates that while there was an overall decrease in informal sector employment in quarter 1 of 2021 (19, 000 jobs), employment in that sector grew by 184,000 jobs in quarter 2 of 2021. Formal sector employment declined by 375,000 jobs in quarter 2 of 2021 compared to quarter 1. Formal sector employment in 2021 quarter 2 grew by 136,000 jobs compared to the previous year, driven primarily by the construction (115,000 jobs), community and social services (42,000 jobs), transport (33,000 jobs), mining (23,000 jobs) and utilities (4,000 jobs) industries.

Despite political, economic, and demographic changes, the prevailing employment arrangement in South Africa can be traced to the era of apartheid and the post-apartheid period. During the apartheid regime in 1970, employment was based on racial disparity and discrimination. At that time, most formal employment opportunities were reserved predominantly for the white population to the exclusion of all other population groups. This practice created a large swath of the population that was unrepresented in the workforce. The policy also created a huge skills gap in the population (Venter and Levy 2014). Toward the end of the 1990s, the post-apartheid era saw the promulgation of far-reaching legislation, programs, and policies to redress those past imbalances.

Recent research in South Africa has shown that economic growth has demanded more skilled workers than unskilled workers. Thus, the policies of the pre-democratic period turned out to have resulted in labor force deficits which the post-1994 democratic government has had to grapple with. These short-comings needed to be addressed alongside considerations for providing decent employments that incorporate fair labor practices that are in line with the ILO principles. In this respect, therefore, the employment situation in South Africa continues to be constrained while efforts are being made to create the right environment for the emergence of adequate workforce that would take advantage of decent employment opportunities in society.

Intersection of Unemployment and Poverty in South Africa

The trend of poverty across African nations, particularly in South Africa, is alarming. The trend can be attributed to unproductive and precarious employment. Poverty in South Africa can be viewed in a broader context as being more widespread among Black people, the uneducated, female-headed households, large families, and children. The prevalence of poverty has been determined to have a strong spatial dimension in South Africa—a demonstration of the enduring effect of apartheid (World Bank 2014).

Following the radical constitutional change in South Africa, the government has been compelled to design a number of programs and policies for reducing poverty; improving access to basic services, education, and healthcare; and providing social protection and economic opportunities. This in turn helped to reverse some of the adverse effects of the system of segregation and deprivation under apartheid. However, this focus is being weakened by the country's low economic growth prospects (World Bank 2014).

South Africa has a high concentration of low-income earners (the "poor") and a few very high-income earners (the "rich" or "elite") but only a small number of middle-income earners. This has resulted in a high level of income polarization (Finn 2015). The "World Bank Group Poverty and Equity Report" (2020) emphasized that 55.5 percent, or roughly 30.3 million people, of the population is living in abject poverty, while about 13.8 million (25 percent) are currently experiencing "sustenance poverty" in South Africa. Correspondingly, poverty is measured at the international poverty line of between $1.90 and $3.20 per person per day (Bello 2006). Because it is often said that "virtuous occupations" are the panacea for poverty reduction, alleviation, and eradication within any society, the impact of South African government policy interventions, such as the Labour Relations Act of 1995, the Employment Equity Act of 1998, and the Basic Condition of Employment Act of 1997, on inequality, and thus on poverty, has been very modest. Creating good jobs for the poor would have a much larger impact in eradicating inequality and poverty.

Poverty is rooted in unemployment and underemployment. The poor are often unable to improve their quality of life without the income they earn through work. To reduce poverty and ensure sustainable economic and social development, it is necessary to create productive ("decent") employment opportunities. Providing decent jobs for women and young people, especially, is crucial for securing income and empowering the poor. An expansion of the economy under circumstances of rapid growth can increase the number of productive and remunerative jobs, which can decrease poverty levels. Although economic growth is important to poverty reduction, it is not the only factor. The distribution of income will be determined by the ability of the poor to respond to the increasing demand for labor in more productive job categories.

Implications of Unemployment and Poverty for Socioeconomic Development

Unemployment and poverty are global phenomena that the government of each must combat. Creating jobs must occupy a central place in national poverty reduction strategies, given the importance of employment in reducing poverty (Mafiri 2002). A number of employment strategies are often connected to agricultural and regional development, such as promoting micro projects in rural areas and using labor-intensive agricultural technologies. In addition to promoting self-employment and nonfarm employment in rural areas, providing targeted employment interventions, microfinance, and credit as resources for skill development and training are some additional strategies that encourage individuals to become self-employed (OECD Economic Surveys 2008).

The problem with such strategies, however, is that they commonly address the quantity of employment while overlooking or underestimating qualitative aspects such as equity, security, dignity and freedom. Generally speaking, national poverty reduction strategies do not address decent employment principles such as social protection or worker rights. Social perspectives on development promote the view that decent work is the most effective means of achieving socioeconomic development, eradicating poverty, and enhancing personal well-being. To achieve international development goals, including the Millennium Development Goal of halving extreme poverty, productive employment opportunities are critical. Better and more productive jobs should be created, especially those that can absorb high concentrations of low-wage workers. A few elements are necessary to create such jobs. They include investing in labor-intensive industries (such as agriculture), promoting a shift in employment structure to more productive sectors and occupations, and improving employment quality in the informal economy. The provision of skills and assets to the poor is also imperative. Consequently, the poor will be better able to capitalize on any expansion in employment possibilities.

Methodology

Nonexperimental descriptive qualitative research was employed in the study. In this analysis, the qualitative method was selected because it enumerated the aspects of precarious employment in relation to poverty. Published reports pertaining to the labor force, employment, and unemployment rates in South Africa were analyzed. Furthermore, the reports of the Quarterly Labour Force Survey of Statistics South Africa were used in the analysis.

Findings and Discussion

Table 1 depicts the labor force of South Africa from April 2020 through June 2021. It should be noted that labor market interaction for 2021 does not encompass the entire economy. It takes into account the population 15 through 64 years old in the labor force—that is, employed people in formal and informal industries, in agricultural production, and in private households, as well as the unemployed and the economically inactive in South Africa. The adult population increased by 145, 000 (0.4 percent) from the first quarter of 2021. An additional 578,000 workers (1.5%) ages 18 through 24 joined the workforce. Thus, the number of employed people decreased by 54,000 in quarter 1 2021, and the number of unemployed had increased by 584,000, resulting in an increase of 530,000 (2.4%) among people who were in the labor force. The number of discouraged job seekers grew (4.1%) between the two quarters, but the number of those not actively searching decreased (4.1%). As a result, 386,000 fewer people were unemployed between the two quarters (Stats SA 2021).

Table 1. South Africa's decent work indicators (adapted from Stats SA 2021)

	Apr–Jun 2020	Jan–Mar 2021	Apr–Jun 2021
Population, 15–64 years old	39,021	39,455	39,599 (0.4%)
Labor force	18,443	22,237	22,768 (2.4%)
Employed	14,148	14,995	14,942 (–0.4%)
Formal sector (nonagricultural)	10,064	10,574	10,200 (–3.5%)
Informal sector (nonagricultural)	2,280	2,502	2,686 (7.4%)
Agriculture	799	792	862 (8.7%)
Private households	1,005	1,127	1,194 (6.0%)
Unemployed	4,295	7,242	7,826 (8.1%)
Not economically active	20,578	17,218	16,832 (–2.2%)
Discouraged work-seekers	2,471	3,131	3,317 (5.9%)
Other(not economically active)	18,107	14,086	13,515 (–4.1%)
Unemployment rate	23.3%	32.6%	34.4%
Employed/population ratio (absorption)	36.3%	38.0%	37.7%
Labor force participation rate	47.3%	56.4%	57.5%

The differences between individual labor market status categories were taken into account in order to better understand the observed large shifts between 2021 quarters 1 and 2. The unemployment rate jumped from 32.6% to 34.4% between the second quarter and the third quarter. This was due to a large number of individuals moving from employable and inactive statuses to unemployed statuses during those quarters. Since the start of the Quarterly Labour Force Survey in 2008, this is the highest unemployment rate ever recorded. As a result of these movements, the labor force participation rate also increased by 1.1 percentage points, to 57.5%, in quarter 2 of 2021 compared to quarter 1. The absorption rate fell by 0.3 percentage points, to 37.7%, in the second quarter of 2021 over the first quarter (Stats SA 2021).

In quarter 2 of 2021, only 375,000 jobs were lost in the formal sector, while 184,000 were added in the informal sector, 69,000 in the agricultural sector, and 67,000 in private households. This is an indication that in 2021 the informal sector received more growth than the formal sector in South Africa. Overall, employment grew by 793,000 jobs since the previous year. There were 3.5 million more unemployed people in the labor force, while there were 3.7 million more people who were not economically active (Stats SA 2021).

The population of South Africa has grown exponentially, but the unemployment rate has risen by 8.1%, and the employment rate has decreased by 0.4%. These numbers have been boosted by the discouragement of work-seekers, a number that has risen 5.9%. This has led to chronic poverty, one of the most intractable challenges for South Africa's workforce despite numerous government interventions. South Africa's Accelerated and Shared Growth Initiative is one of many policies enacted by government to promote decent employment and to promote economic growth. The policy framework's main objective is to create jobs. According to the plan, a million jobs would have been created by this program in the five years prior to 2020. This combination of factors, coupled with the current stagnant economic growth in South Africa, makes it vitally important for funders, policy makers, and those working on projects aimed at employment creation to evaluate and invest in programs that promote positive attitudes toward employment.

Conclusion

It can be deduced from the above analysis that there are many intricacies within the South African labor market, especially in the employment creation environment. The employment rate has dropped dramatically, while the unemployment rate continues to rise, which was happening even before the COVID-19 pandemic. It is feared that if the experience of the pandemic is prolonged, it will continue to negatively affect the decent employment situation in South Africa and around the African continent and the world. The analysis thus far points to the trend that the informal sector has grown more nominally than the formal sector in terms of employment in South Africa. This trend does not, however demonstrate that decent employment opportunities have been created in the economy.

One thing that the trend does show though is that large-scale unemployment remains a major socioeconomic challenge to South Africa. This is more so because it has been established that unemployment or indecent employment breeds poverty in society.

Recommendations

- Decent employment that is in sync with ILO principles can help society to combat the challenge of poverty.
- A well-considered policy of government to empower the informal sector can help in the employment creation efforts of the country.
- There is a need to redouble efforts to ensure that the underpinning considerations that led to policy enactments to redress historical imbalances and inequalities of the South African employment environment is concertedly pursued.
- The provision of decent employment should be vigorously pursued by government in order to eradicate poverty.

References

Altman, M.A. (2002). Employment policy in a minerals economy. *Journal of International Development*, 13(6), 681–709.

Bell, S. and Newitt, K. (2010). Decent work and poverty eradication: Literature review and two-country study. London: Ergon Associates.

Bello, M. (2006). "Reflections on poverty reduction strategies" in A. Saliu et al., eds., *Democracy and Development in Nigeria. Vol.: 2 Economic and Environmental Issues*. Lagos: Concept Publishers.

Cohen, T., and Moodley, L. (2012). Achieving "decent work" in South Africa. *Potchefstroom Electronic Law Journal*, 15(2), 319–344.

Deranty, J.-P. and MacMillan, C. (2012). The ILO's decent work initiative: Suggestions for an extension of the notion of "decent work". *Journal of Social Philosophy*, 43 (4), 386–405.

Edwards, L. (2001). Globalisation and the skills bias of occupational employment in South Africa. *South African Journal of Economics*, 69(1), 40–71.

Finn, A. (2015). A national minimum wage in the context of the South African labour market. Working Paper Series No. 1, University of Witwatersrand. https://nationalminimumwage.co.za/wp-content/uploads/2015/09/NMW-RI-Descriptive-Statistics-Final.pdf

ILO (1999). Decent work. Report of the Director-General, International Labour Conference 87th Session. Geneva: ILO.

ILO (2021) Advancing social justice, promoting decent work. ILO/decent work. https://www.ilo.org/jakarta/whatwedo/publications/WCMS_714338/lang--en/index.htm

Isaacs, G. (2016). A national minimum wage for South Africa. National Minimum Wage Research Initiative, Summary Report No. 1. Johannesburg: University of the Witwatersrand.

Johannes. F. (2012). The cost of rigidity: The case of the South African labour market. *Comparative Economic Studies*, 54, 809–842.

Mafiri, M.I. (2002). Socio-economic impact of unemployment. Master's thesis, University of Pretoria. https://repository.up.ac.za/bitstream/handle/2263/27288/Complete.pdf?sequence=3

Meth, C. (2001). Unemployment in South Africa—What the latest figures tell us. Unpublished paper, University of Natal.

Narayan, D., Chambers, R., Shah, M.K., and Petesch, P. (2000). *Voices of the Poor: Crying Out for Change*. New York: Oxford University Press for the World Bank.

OECD Economic Surveys. (2008). South Africa: Economic Assessment 2008, Vol. 2008/15. Paris: OECD Publications.

Paul, K.I., and Moser, K. (2009). Unemployment impairs mental health: Meta-analyses. *Journal of Vocational Behaviour* 74, 264–282.

Stats SA (2021). Stats South Africa, 2021 Quarterly Labour Force Survey: Quarter 1. www.statssa.gov.za/publications/P0211/P02111stQuarter2021.pdf

Swanson, J.L. (2012) "Work and psychological health" in N.A. Fouad, J.A. Carter, and L.M. Subic, eds., *APA Handbook of Counseling Psychology*. Washington, DC: American Psychological Association), 3–27.

Venter, R., and Levy, A. (2014). *Labour Relations in South Africa*. Oxford: Oxford University Press.

Wilson, W.J. (1996). *When Work Disappears: The World of the New Urban Poor*. New York: Knopf.

World Bank. (2014). South Africa economic update: Fiscal policy and redistribution in an unequal society. Washington, DC: World Bank.

World Bank Group (2020). World Bank Group poverty and equity report: Sub-Saharan Africa, poverty data. World Bank. www.worldbank.org/poverty

Gender and LGB Pay Gaps in the National Health Service

KAREN MUMFORD
University of York and IZA

EDITH AGUIRRE
ISER, University of Essex

ANNA EINARSDÓTTIR
University of York

BRIDGET LOCKYER
Bradford Institute for Health Research

MELISA SAYLI
University of Surrey

BENJAMIN A. SMITH
Rotherham NSH Foundation Trust

Abstract

Studies of the relationship between sexual orientation and pay have faced difficulties applying standard models of discrimination if orientation is not observable. Analogously, behavioral explanations of pay based on models of gender-linked within-household specialization may not be as relevant in a nonheterosexual context. This article explores pay gaps in the English National Health Service (NHS) using information including earnings, gender, LGB (lesbian, gay, bisexual) identity, coupling status, and the disclosure of sexual orientation. The results reveal a robust gender pay gap of 4 percent in favor of males, but no overall LGB pay gap compared to heterosexuals. The latter is found to be due to similar-sized offsetting effects from disclosure on LGB pay relative to comparable heterosexuals. Considerable heterogeneity was established among the LGB employees, with disclosure associated with 13 percent more pay, with three-quarters of this gap related to unexplained differences in returns to observable characteristics. The finding that disclosure of sexual orientation is related to more favorable pay treatment for LGB employees is true for both male and female employees within the NHS.

Introduction

There are a vast number of studies exploring gender pay gaps in the economics literature (Blau and Kahn 2017). Studies that also address potential implications of LGBTQ+ status for pay gaps are more recent and considerably rarer (Badgett, Carpenter, and Sasone 2021). Using a rich new survey of National Health Service (NHS) employees in England, this article seeks to provide a more complete explanation of pay gaps by including information on gender, LGB identity, coupling status, and disclosure of sexual orientation within workplaces. To the best of our knowledge, this is the first study to include direct measures of both disclosure of LGB sexual orientation and coupling status, allowing for a more insightful interpretation of the mechanisms behind gender and LGB pay gaps.

The most commonly used model explaining wage outcomes, and pay gaps, is the human capital model (HCM). In its simplest form, the HCM argues that wages increase with investments made in the productivity of the individual, especially investments in formal education and on-the-job training (Becker 1975; Mincer 1974). The HCM rapidly evolved from an intrinsically individual perspective to a broader household context (Becker 1985). Individuals who expect to eventually become members of a household may also expect to specialize in different tasks within the household to maximize the combined utility of the household's members. For example, if women are expecting to spend time out of the labor market to raise children, they may invest less in formal labor market skills and/or choose to enter occupations that require less on-the-job training, thereby lowering their predicted earning capacity. Employers may also expect that women will be less attached to the labor market and will have shorter job tenure. The outcome can become self-fulfilling if employers deny women hiring opportunities and/or training paths associated with longer tenure and higher pay.

With such an observable physical characteristic as gender, it may be difficult for women to avoid this lower pay outcome. One way could be for women to engage in expensive formal education as a signal to employers that they intend to stay in the labor market to reap the returns of the investment (Spence 1973). This will be a higher-risk investment for women than men if some employers simply have a taste for discrimination and deny women opportunities regardless of qualifications (Becker 1957) or if employers do not recognize the individual has different aspirations from their group average (Arrow 1973; Phelps 1972).

LGB (lesbian, gay, bisexual) employees may be making similar decisions regarding the allocation of market and nonmarket work within households. They may also be facing employers with a taste for discrimination (Badgett 1996). However, identifying as LGB is a nonobservable characteristic in the workplace unless the employee chooses to share this information.

There is a small but influential body of literature on LGB versus heterosexual pay gaps based on analysis of survey respondents providing individual-level information on their pay, cohabitation, and sexual orientation. Unfortunately, such studies are often constrained by low numbers of LGB respondents. It is also very rare for the analyst to have information on disclosure of sexual orientation and the relative pay of employees.

In her seminal study, Badgett (1995) found that gay men (or lesbians) living in same-sex relationships in the United States earn less than do comparable heterosexuals living in different-sex partnerships. Klawitter (2015) used meta-analysis to conclude that most studies find lesbians typically earn more than comparable heterosexual women and that the negative male gay pay gap is diminishing over time. Jepsen and Jepsen (2020) argued that the male gay pay gap diminished in the United States until 2010, flattening thereafter, with the positive lesbian pay gap continuing to decrease but at a very slow rate. More recently, Badgett, Carpenter, and Sason (2021) found that the negative pay gap for gay/bisexual men has declined in the United States since 2000 but not in a consistent manner and appears to be showing persistence at around 7 log percentage points (lpp) since 2015, while the wage advantage for lesbian/bisexual women has been falling since 2000 (when it was around 10 lpp) and is now close to zero. These findings are arguably consistent with behavioral models of within-household specialization, with gay men being less labor market–oriented than heterosexual "primary household-earner" males, and lesbian women being more labor market–oriented than heterosexual "secondary household-earner" women (Aksoy et al. 2018; Aksoy et al., 2019; Jepsen and Jepsen 2020). It is not possible, however, in those studies to separate out potential discrimination effects because they do not include information on whether sexual orientation is identifiable in the workplace.

Arabsheibani, Marin, and Wadsworth (2005) showed that it is important to control for household structure when comparing relative wages. Using UK data, they found that gay men earn more than heterosexual men but that a statistically significant pay differential only occurs between cohabiting gay men and heterosexual men who live with a female partner. They found that this pay gap would be twice as large (from 2.5 to 5 lpp) if the characteristics of gay men were rewarded equivalently to those of heterosexuals. In contrast, Aksoy et al. (2018) found a small pay penalty for partnered gay men relative to partnered heterosexual men in the United Kingdom. They posited that larger pay penalties for older and unpartnered gay men may be related to their long-term lack of a heterosexual marriage making their sexual orientation more observable to their work colleagues, thereby opening the door to discrimination. Bridges and Mann (2019) sought to further address this issue by using information from the UK Labour Force Survey on same-sex legal partnership as an indicator of how open gay men or lesbians are in their workplaces, arguing that those who have made a legal commitment to their same-sex relationship are more likely to be open with their colleagues about their orientation. They find that, while both gays and lesbians earn more than equivalent-gender heterosexuals, this form of disclosure is associated with a lower pay premium and with less promotion for gay men relative to male heterosexuals. They argued that these findings are consistent with negative discrimination toward recognizable GB status for males in UK workplaces.

In a very rare study, Plug and Berkhout (2008) combined information on earnings, sexual orientation, and disclosure to consider pay gaps. They used data on young Dutch males, two years post-college graduation, who work full-time and are not self-employed. Plug and Berkhout (2008) found that these gay men earn 3 to 4 lpp less pay than do the heterosexual males; they argued that this pay gap is driven by undisclosed gay men concentrating in lower-paid, less-productive occupations, where they earn some 5 to 9 lpp less than other men. While Plug and Berkhout (2008: 10) rejected a positive discrimination option, they did also find a positive pay return of some 3 to 8 log percentage points associated with disclosure. Plug and Berkhout (2008) additionally provided a very useful discussion on the possible endogeneity between pay and disclosure, the inherent difficulties in separating out discrimination and productivity differences using cross-sectional data, and the inability to locate a meaningful identifying variable for the disclosure decision in empirical analysis. These are all issues that are relevant to our study, and we will return to consider the related caveats they impose below.

The next section of our article describes the NHS data source and provides information on the variables to be included in the modelling. This is followed with discussion of the methodology and explanation of the empirical estimation of the earnings function. Results from this estimation are presented with a range of comparable pay gaps. Decomposition analysis is subsequently presented to provide greater insight into the components of the pay gaps. Disclosure is found to have a substantial association with pay differences between the LGB and with comparable heterosexual employees. Further discussion and conclusions are presented in the final section.

Data

Data Collection

There are very few data sets that include information on sexual orientation (Hudson-Sharp and Metcalf 2016). To the best of our knowledge, there are no data sets that include information of the determinants of pay, partnership, and disclosure of sexual orientation in the workplace. In response to this lack of relevant information, the project team developed an online survey—the Employee Engagement Survey (EES-NHS)—of those employees working in NHS trusts in England who are covered by the NHS Pay Review Board (NHSPRB).[1] Full information on the surveying procedure and sample characteristics are provided in the survey technical report (Einarsdóttir et al. 2020). The NHS is a particularly relevant workforce to survey because it is large enough to generate a suitable LGB sample for statistically meaningful analysis. Furthermore, the NHS employees included are all working in the health sector, where they share a common employer, with well-recognized pay and working conditions set by the NHSPRB (which also means that there are no doctors or dentists in the sample). The NHS has a reputation for being an employer mindful of discrimination and with a varied (in terms of nationality, ethnicity, gender, and/or sexual orientation) and highly unionized workforce (Einarsdóttir et al. 2020). These commonalities help to focus the empirical analysis presented below, however; they may also limit legitimate extrapolation of the findings outside of the NHS to other less-supportive workforces in England (Bryson 2017; Wang, Gunderson, and Wicks 2018; Wax, Coletti, and Ogaz2018). This is a caveat that will be returned to in the discussion and interpretation of the results below.

The EES-NHS was launched in January 2019 and closed in May 2019. Human resources representatives and equality, diversity, and inclusion leads were approached in all 226 trusts in NHS England and asked to disseminate the survey to employees in their trust. Eleven trusts did not respond to the EES-NHS, and a further three declined to participate because of survey fatigue. During the survey dissemination process, the project team worked in partnership with NHS employers and the NHS confederation to promote the survey. This included a communications campaign package from NHS employers, led by the NHS confederation's communications team. NHS Digital's headcount data from September 2018[2] suggests that the potential sample frame was 1.19 million (staff working in NHS trusts in England), implying a response rate of less than 1 percent for the EES-NHS. Such a low response rate raises obvious concerns that the sample does not reflect the population of NHS employees.

Compared to the 2018 NHS Staff Survey (NHS-SS), the EES-NHS sample had a similar gender breakdown (with around 77 percent female employees) and age distribution (Table 8 in Einarsdóttir et al. 2020). In terms of sexual orientation, however, the EES-NHS sample has a larger proportion of respondents declaring as LGB (12 percent compared to 3.5 percent) but fewer respondents opting for "prefer not to say" (2.3 percent relative to 6.5 percent). This is not surprising given LGB labeling included in the advertising for the EES-NHS survey; it may also indicate potential fear of a backlash for disclosing minority sexual identity in the NHS-SS.

An overall comparison of the EES-NHS with the 2018 NHS-SS showed that many of the survey items displayed similar patterns. The degree of openness about sexual orientation with all or most co-workers in the EES-NHS sample (60.3 percent) also followed a similar pattern as in the national LGBT survey (62.5 percent). Similarly, 78.2 percent of respondents in the national LGBT survey subsample were satisfied with their lives,[3] but 70.9 percent of the EES-NHS respondents are. See Einarsdóttir et al. (2020) for more detailed comparison across these data sets.

The full sample taken from the EES-NHS included 3,724 of NHS employees. Missing observations for variables used in the analysis limited the usable sample to 3,556 observations, (12.37 percent LGB). One compensation for the over-representation of LGB employees in the EES-NHS survey was the inclusion of a reasonable number of LGB observations in the analysis (440 LGB). Nevertheless, the EES-NHS sample size was not big, and the sampling process was not random—both of which limit the extrapolation of the findings across the full NHS workforce.

Variable definitions and summary statistics are presented in Table 1 (starting on the next page): pairwise statistical testing for differences in mean values are included for the male and female samples (columns 2 and 3), the LGB and heterosexual (columns 4 and 5), male GB and male heterosexual (columns 6 and 7), and female LB and female heterosexual (columns 8 and 9). Fuller variable definitions and further summary statistics are provided in the appendix (Table A1).

The Structure of Pay in the NHS and Other Explanatory Variables

The employees in the NHS sample considered in this article are paid in bands, and those bands are set by the government with the advice of the NHSPRB.[4] The average hourly wage measure used below is constructed from the midpoint of the employee's salary band, allowing for their usual working hours and adjusting for paid overtime hours. On average, the employees in the sample receive a salary of £16.62 per hour (column 1 of Table 1)—£17.36 for males (column 2) and £16.42 for females (columns 3)—suggesting a statistically significant raw gender pay gap of 4.3 percent or 4.4 lpp at the 95 percent confidence level.[5]

Survey respondents were categorized as LGB according to their own choices. The LGB workers who responded that they are open about their sexual orientation with their supervisor/manager and co-workers in their current job were counted as disclosed. The gender of choice from the survey respondents was also used, which is clearly relevant for transgender respondents.[6] On average, LGB workers received £16.83 per hour (column 4), 1.3 lpp more than heterosexuals, at £16.59 per hour (column 5), although this raw pay gap is not statistically significant. We also found no statistically significant ethnic pay gap in column 1, although within the ethnic minority group, males earned more than females, and heterosexual males earned more than GB males.

Table 1. Means (and standard deviations) of variables by sexual orientation

						Male		Female	
	All	**Male**	**Female**	**LGB**	**HetS**	**GB**	**HetS**	**LB**	**HetS**
	(1)	**(2)**	**(3)**	**(4)**	**(5)**	**(6)**	**(7)**	**(8)**	**(9)**
salary	16.62	***17.36***	***16.42***	16.83	16.59	17.71	17.23	16.03	16.46
	(6.53)	(7.27)	(6.30)	(6.80)	(6.49)	(7.61)	(7.14)	(5.86)	(6.34)
natural log salary	2.742	***2.777***	***2.733***	2.754	2.741	2.80	2.77	2.72	2.73
	(0.36)	(0.39)	(0.35)	(0.36)	(0.36)	(0.39)	(0.39)	(0.33)	(0.36)
LGB and disclose						0.60		0.42	
Demographics									
male	0.21			***0.48***	***0.17***				
LGB	0.12	***0.28***	***0.08***						
age	46.21	***45.03***	***46.53***	***41.46***	***46.88***	***41.89***	***46.24***	***41.06***	***47.01***
	(11.43)	(11.79)	(11.31)	(11.32)	(11.28)	(11.23)	(11.78)	(11.40)	(11.17)
ethnic minority	0.11	***0.14***	***0.10***	**0.09**	**0.12**	***0.10***	***0.16***	0.08	0.11
married	0.51	0.49	0.51	***0.29***	***0.54***	***0.27***	***0.57***	***0.30***	***0.53***
live in couples	0.69	0.70	0.68	***0.57***	***0.70***	***0.56***	***0.75***	***0.59***	***0.69***
dependent children	0.32	0.30	0.32	***0.14***	***0.34***	***0.07***	***0.39***	***0.21***	***0.33***
disabled	0.36	0.37	0.35	***0.45***	***0.34***	0.39	0.36	***0.50***	***0.34***
caregiver responsibilities	0.26	***0.19***	***0.28***	**0.23**	**0.27**	0.19	0.19	0.27	0.28
foreign born	0.12	**0.14**	**0.12**	***0.09***	***0.13***	0.11	0.15	***0.07***	***0.12***
Qualifications									
Min qual	0.01	0.01	0.01		0.01		0.01		0.01
GCSE, D-G	0.05	0.04	0.05	***0.02***	***0.05***	0.03	0.04	**0.02**	**0.05**
GCSE, A-C	0.08	**0.07**	**0.09**	***0.05***	***0.09***	0.05	0.07	***0.05***	***0.09***
trade	0.004	***0.01***	***0.002***	0.002	0.004		0.01	0.01	0.01
A levels	0.09	0.10	0.09	0.10	0.09	0.10	0.10	0.10	0.09
HE and TQ	0.16	0.17	0.16	0.15	0.16	0.16	0.17	0.14	0.16
first degree	0.30	0.31	0.30	0.32	0.30	0.29	0.32	0.34	0.29
higher degree	0.28	0.27	0.28	***0.32***	***0.27***	***0.34***	***0.24***	0.30	0.28
potential work experience	17.96	***16.02***	***18.48***	***15.12***	***18.36***	**14.92**	**16.45**	***15.29***	***18.76***
	(11.66)	(11.11)	(11.75)	(10.57)	(11.75)	(10.24)	(11.42)	(10.87)	(11.78)

					Male		Female	
	All	**Male**	**Female**	**LGB**	**HetS**	**GB**	**HetS**	**LB**	**HetS**
	(1)	**(2)**	**(3)**	**(4)**	**(5)**	**(6)**	**(7)**	**(8)**	**(9)**
Occupation									
allied	0.19	0.19	0.19	0.21	0.19	0.17	0.20	***0.24***	***0.19***
ambulance	0.009	***0.03***	***0.003***	***0.04***	***0.004***	***0.06***	***0.01***	***0.03***	***0.01***
public health	0.01	**0.01**	**0.009**	0.01	0.01	0.02	0.01	0.01	0.01
commissioning manager	0.01	***0.02***	***0.01***	0.02	0.01	0.01	0.03	0.02	0.01
nurses	0.24	***0.14***	***0.27***	**0.21**	**0.25**	***0.18***	***0.12***	0.23	0.27
nursing auxiliary	0.05	0.05	0.05	0.06	0.05	0.07	0.04	0.06	0.05
social care	0.006	0.007	0.006	0.004	0.007	0.01	0.01	0.01	0.01
wider	0.24	***0.21***	***0.25***	***0.19***	***0.25***	0.18	0.22	**0.20**	**0.25**
general management	0.09	***0.14***	***0.08***	0.11	0.09	0.15	0.14	0.07	0.08
other	0.10	***0.15***	***0.09***	0.10	0.10	***0.11***	***0.17***	0.10	0.09
health professional	0.44	***0.35***	***0.46***	0.43	0.44	**0.40**	**0.33**	0.45	0.46
Job Characteristics									
part-time	0.24	***0.09***	***0.28***	***0.11***	***0.26***	0.06	0.10	***0.15***	***0.29***
job permanent	0.93	0.92	0.93	0.93	0.93	0.94	0.92	0.92	0.93
trade union member	0.57	***0.52***	***0.58***	0.56	0.57	0.57	0.50	0.56	0.59
tenure	6.87	***6.20***	***7.05***	***5.53***	***7.06***	**5.48**	**6.48**	***5.58***	***7.18***
Workplace Characteristics									
mentor	0.47	***0.42***	***0.48***	0.45	0.47	0.43	0.42	0.48	0.48
happy training	0.47	0.45	0.47	0.48	0.47	**0.50**	**0.43**	0.47	0.47
friend	0.61	***0.49***	***0.64***	***0.56***	***0.61***	0.54	0.48	**0.59**	**0.64**
cooperative	0.39	0.41	0.39	0.43	0.39	0.44	0.40	0.42	0.39
bully witnessed	1.84	1.86	1.84	1.83	1.84	1.91	1.83	1.76	1.85
discrimination	1.95	1.95	1.95	1.97	1.95	1.93	1.96	**2.01**	**1.95**
responsive hours	0.46	***0.39***	***0.48***	0.46	0.46	0.39	0.40	0.52	0.47
pressure	0.55	0.55	0.56	0.55	0.56	0.50	0.56	0.58	0.55
co-worker support	0.77	0.75	0.78	0.80	0.77	0.76	0.74	***0.83***	***0.77***
work–life balance	0.59	0.59	0.59	0.61	0.59	**0.64**	**0.57**	0.58	0.59
supervisor support	0.61	0.59	0.62	0.60	0.61	0.58	0.59	0.62	0.62

						Male		Female	
	All	**Male**	**Female**	**LGB**	**HetS**	**GB**	**HetS**	**LB**	**HetS**
	(1)	**(2)**	**(3)**	**(4)**	**(5)**	**(6)**	**(7)**	**(8)**	**(9)**
NHS England Region									
North of England	0.23	0.23	0.23	0.23	0.23	**0.19**	**0.24**	0.27	0.23
Midlands and East of England	0.33	***0.28***	***0.34***	***0.25***	***0.34***	**0.23**	**0.29**	***0.26***	***0.35***
London	0.15	***0.17***	***0.14***	***0.24***	***0.14***	***0.27***	***0.14***	***0.21***	***0.14***
South West	0.11	0.12	0.11	***0.07***	***0.12***	***0.07***	***0.13***	***0.06***	***0.12***
South East	0.15	***0.18***	***0.15***	***0.19***	***0.15***	**0.22**	**0.17**	0.16	0.15
Trust Type									
Acute Specialist Trusts	0.02	***0.03***	***0.01***	***0.05***	***0.01***	0.05	0.03	***0.04***	***0.01***
Acute Trusts	0.50	0.51	0.50	***0.38***	***0.51***	***0.41***	***0.55***	***0.36***	***0.51***
Ambulance Trusts	0.01	***0.03***	***0.008***	***0.04***	***0.008***	***0.06***	***0.01***	***0.03***	***0.01***
Combined Acute And Community Trusts	0.12	0.12	0.12	0.11	0.12	0.11	0.13	0.10	0.12
Combined Mental Health/ Learning Disability And Community Trusts	0.08	**0.07**	**0.09**	***0.11***	***0.08***	***0.10***	***0.05***	0.11	0.08
Community Trusts	0.10	***0.05***	***0.11***	**0.08**	**0.10**	0.07	0.05	0.09	0.11
Mental Health/Learning Disability Trusts	0.14	0.15	0.13	***0.20***	***0.13***	0.17	0.15	***0.23***	***0.13***
Observations	3,556	753	2,803	440	3,116	210	543	230	2,573

Mean pair differences: males (2) vs. females (3); LGB (3) vs. heterosexual (4); among males: GB (6) vs. male heterosexual (7); or among females: LB (8) vs. female heterosexual (9).

bold p < 0.10, ***bold and italic p < 0.05***

According to the HCM, it is reasonable to expect wages to increase with measures related to investments made in the productivity of the individual, especially their education, job training, and work experience (Becker 1975; Mincer 1974). The earnings function estimated below is augmented with the addition of further categories of explanatory variables, including demographic variables which may affect an individual's choice of jobs (gender, LGB identity, having dependent children, marital status, ethnic identification, being foreign born, being disabled, being a caregiver, and age); occupation controls; job characteristics that are a range of variables loosely reflecting the individual's response to the labor market (working part-time, having a permanent contract, current job tenure, and being a trade union member); workplace characteristics that are associated with the workplace but can vary across employees within that work location (having an effective mentor, having supportive co-workers, a friend in the workplace, being happy with training opportunities, being able to use responsive working hours, often feeling under pressure, ability to maintain work–life balance, having a supportive supervisor, and being in a cooperative work place); and trust controls that are common to all workers in that trust (regional location, and trust type). This augmented model is referred to below as the full model.

The great majority of the sample being considered is female (some 79 percent). Compared to the males, these women are on average older, have more work experience, are twice as likely to be nurses, half as likely to be in general management, and three times more likely to work part-time (Table 1). The women are also more likely to have a mentor, belong to a trade union, have at least one close friend in their workplace, and make use of responsive work-hour provisions.

Sexual orientation is not evenly distributed across the genders in the sample: 28 percent of the men identify as GB (column 2 of Table 1) and 8 percent of the women as LB (column 3). Disclosure of this orientation in the workplace is more common among men than women (60 percent relative to 42 percent; see columns 6 and 8).

Columns 6 and 7, and 8 and 9, of Table 1 reveal that LGB workers in the study were on average younger than equivalent-gender heterosexuals, and they are less likely to be from an ethnic minority, married, be living with their partner, or have dependent children. They have less work experience on average, tend to have higher education (especially the males), are more likely to work as nurses if male, less likely to work part-time if female, less likely to work in acute trusts, and more likely to work in ambulance trusts than heterosexuals. Formal estimations of pay gaps within and between these groups are considered next.

Estimating the Earnings Functions

Following the literature, examining the wage differentials developed by Becker (1975) and Mincer (1974), and using semi-logarithmic wage equations, the earnings equation was estimated as follows:

$$W_{il} = X'_{il}\beta_l + \varepsilon_l, E(\varepsilon_l) = 0, l \in (a, b, p) \quad (1)$$

where W_i is the natural log of the average hourly wage, W, for individual i in group type l; X_i is a vector of explanatory variables and a constant; ε_i is a residual term; and a represents comparison group a; b the alternative comparison group b; or p the pooled group of a and b combined (Neumark 1988). For example, a might be males, b might be females, and p would be all the males and females combined; or a could be set as male GB, b as male heterosexuals, and p would be all the males. An indicator variable identifying group membership is also included in the pooled model. Estimating the earnings function using ordinary least squares and allowing for clustering at the trust level throughout, the first regression specification is a parsimonious model including only indicator variables for gender (male) and being LGB (column 1, panel a, of Table 2). With no additional explanatory variables in the model, men earned 4.4 lpp more than women, and there is neither a sizeable nor a statistically significant pay differential between LGB and heterosexual workers in the study population.

Table 2. Determinants of log earnings (OLS)

ln(salary)	LGB (1)	Base (2)	Full model (3)
(a) Full sample			
Male	0.044*** (0.014)	0.040*** (0.014)	0.0389*** (0.011)
LGB	–0.0001 (0.0193)		
No disclose and LGB		–0.061** (0.024)	–0.049*** (0.017)
Disclose and LGB		0.061** (0.026)	0.044** (0.017)
Adjusted R^2	0.001	0.005	0.618
Number observations	3556		
(b) Male sample			
GB	0.026 (0.028)		
No disclose and GB		–0.081** (0.036)	–0.025 (0.028)
Disclose and GB		0.095*** (0.036)	0.072*** (0.027)
Adjusted R^2	–0.0004	0.0122	0.573
Number observations	753		
(c) Female sample			
LB	–0.018 (0.026)		
No disclose and LB		–0.048 (0.029)	–0.050** (0.019)
Disclose and LB		0.021 (0.038)	0.039* (0.023)
Adjusted R^2	–0.0001	0.0002	0.637
Number observations	2803		

Standard errors in parentheses (clustered at trust level).

*p < 0.10, **p < 0.05, ***p < 0.01.

In addition to the coefficients listed, the full model (column 3) includes additional explanatory and control variables as defined in the data discussion. Complete results are provided in the appendix tables.

In column 2 of Table 2, the LGB respondents were divided into those who have disclosed their sexual identity in the workplace and those who have not, and these two groups were compared separately to the omitted heterosexual category. The LGB workers who disclosed their sexual orientation earned 6.1 lpp more than heterosexuals, and the LGB workers who did not disclose earned 6.1 lpp less than heterosexuals. The two equally sized effects cancel out an overall LGB pay effect (as shown in column 1).

The full set of explanatory variables as defined above (see also Table 1) were included in the model, and selected results (for gender and LGB disclosure) are reported in column 3.[7] The goodness-of fit-measure (adjusted R^2) suggests this full model captured a reasonable 62 percent of the variation in earnings. There was a slightly smaller gender pay gap at 3.9 lpp and offsetting disclosure effects for LGB respondents of –4.9 lpp without disclosure, and +4.4 lpp for those who do disclose, compared to heterosexuals. There was not a sizeable or statistically significant ethnic pay gap found in any of the models considered.[8]

Similar results are provided for the male and female samples in panels b and c of Table 2. Among the males, the GB respondents who disclosed had a higher pay premium relative to heterosexuals than the pay penalty associated with those who don't disclose (column 2), leading to a statistically insignificant pay gap in favor of GB males of 2.6 lpp (column 1). Our results differed from those of Plug and Berkhout (2008), who found a net pay penalty for GB workers among young Dutch men; however, it was interesting they also found a positive pay premium associated with disclosure. For female NHS employees, the positive disclosure effect was slightly outweighed by the negative nondisclosure, and the overall LB worker pay gap was statistically insignificant at –1.8 lpp.[9] We also found that the pay premiums associated with LGB status and disclosure reported in Table 2 are robust to alternative measures of marital status. A substantial coupling premium occurred for both men and women, and this coupling premium was considerably higher for men (see Table A4), as is well established in the literature. Once disclosure is allowed for, however, there was no significant difference in the pay premiums for LGB respondents who are formally married or living with a partner.[10] This finding not surprising because marital status was being used in earlier studies as a proxy for workplace disclosure of sexual orientation.

Decomposing the Earnings Gaps

Further insight into these pay gaps can be provided via decomposition analysis (Fortin, Lemieux, and Firpo 2011; Oaxaca 1973). Following Jann (2008), the approach adopted to apportion the gap in the mean earnings between groups here is discussed in Oaxaca and Ransom (1994), where the reference set of parameters is given by the pooled estimates, $\hat{\beta}$ (reported in Table A5, column 1). The decomposition of the mean earnings gap was calculated as follows:

$$\overline{W}_a - \overline{W}_b = \{(\overline{X}_a - \overline{X}_b)\}'\hat{\beta} + \{(\overline{X}_a'(\hat{\beta}_a - \hat{\beta}) + \overline{X}_b'(\hat{\beta} - \hat{\beta}_b)\} \qquad (2)$$

where an overbar denotes the mean value. The first component $\{(\overline{X}_a - \overline{X}_b)\}'\hat{\beta}$ is often referred to as the endowment (or explained component) reflecting differences in the averages of the observed characteristics across the groups. The second component $\{(\overline{X}_a'(\hat{\beta}_a - \hat{\beta}) + \overline{X}_b'(\hat{\beta} - \hat{\beta}_b)\}$ is the remaining portion of the gap, which is usually referred to as unexplained (or sometimes as the discrimination component), capturing the sum of the differences in the returns to the two groups.[11] Separating the pay gaps into explained and unexplained components also helps to further address the confounding of productivity and discrimination (positive or negative) raised by Plug and Berkhout (2008).

Aggregate decompositions for the earnings function are presented in Table 3, with each row summarizing a separate decomposition. In row a, the total gender earnings gap is 4.4 lpp in favor of males (column 1). The small but statistically insignificant endowment component (column 2) indicates that on average women have more observable characteristics associated with higher pay than do men. More than all the total gap is, however, associated with men receiving higher returns to their observed characteristics (as shown by the unexplained component of 5.08 lpp in column 3).

No statistically significant pay gap was found between heterosexual and LGB employees (row b), nor between female heterosexuals and LB women (row c). In both cases, the explained and unexplained components of the decomposition were insubstantial and statistically insignificant. Among the males (row d), there was some evidence that heterosexuals are receiving higher returns than GB men for their observed characteristics (the unexplained component in 4.77 lpp); however, the overall pay gap of 2.58 lpp between male heterosexuals and male GBs is not statistically significant.

In contrast, the disclosure results in rows (e) through (g) of Table 3 are striking. The LGB workers who have disclosed their sexual orientation in the workplace (row e) earn on average 12.99 lpp more than those who have not, with 78 percent (10.14 lpp) of this gap related to those who disclosed having higher returns (the unexplained component). Among the males (row g), this result is even stronger: the male GBs who have disclosed their sexual orientation in the workplace had on average 17.59 lpp more pay than the male GBs who have not disclosed, with 10.74 lpp (61 percent) being due to those disclosing being treated more favorably (unexplained component) and 6.84 lpp (39 percent) associated with them having characteristics associated with more productivity and higher pay (endowment component). Among the women (row f), the LB workers who have disclosed also receive higher returns (unexplained), but this is partly offset by the nondisclosing LB women having fewer observable characteristics associated with higher pay; this overall pay gap is smaller at 6.87 lpp and is not statistically significant. Nevertheless, in each case (rows e through g), disclosure was found to be significantly associated with sizeable pay gains related to preferential rates of return for those who disclose. In summary, LGB employees who have disclosed their sexual identity in the workplace receive higher pay, consistent with their receiving more favorable pay rewards given their endowments (sometimes referred to as positive discrimination).

Table 3. Decompositions of the wage gaps, full model

	Wage gap	Explained	Unexplained
	(1)	(2)	(3)
(a) Males vs. females	–0.04402***	0.0067	–0.0508***
	(0.0156)	(0.0112)	(0.0123)
(b) Heterosexual vs. LGB	–0.0132	0.0013	–0.01447
	(0.0195)	(0.0146)	(0.0132)
(c) Females	0.0187	0.0110	0.0077
Heterosexual vs. LB	(0.0250)	(0.0191)	(0.0155)
(d) Males	–0.0258	0.0219	–0.0477**
Heterosexual vs. GB	(0.0287)	(0.0226)	(0.0235)
(e) Nonheterosexual (LGB)	–0.1299***	–0.0285	–0.1014***
Disclosed vs. nondisclosed	(0.0319)	(0.0242)	(0.0253)
(f) Female nonheterosexual (LB)	–0.0687	0.0189	–0.0875**
Disclosed vs. nondisclosed	(0.0432)	(0.0334)	(0.0379)
(g) Male nonheterosexual (GB)	–0.1759***	–0.0684*	–0.1074***
Disclosed vs. nondisclosed	(0.0482)	(0.0398)	(0.0353)

It is not possible with a single cross-sectional data set to address causality between disclosure and pay; individuals may have disclosed before, during, or after their pay changes. A further related concern, as discussed above, is that variables influencing pay may also be influencing disclosure in the analyses. Unfortunately, it has not been possible to locate suitable identifying variables to address this potential endogeneity in the data set. The results, and their implications, need to be interpreted with these caveats in mind.

Discussion and Concluding Comments

Researchers examining the relationship between sexual orientation and pay face difficulties applying standard models of discrimination if orientation is not observable. Analogously, behavioral explanations of pay based on models of gender-linked within-household specialization may not be as relevant in a nonheterosexual context. Using a rich new survey of employees from the NHS in England, information including LGB identity, coupling status, and disclosure of sexual orientation to work colleagues was used to explore pay gaps in this study.

Men were found to earn some 4 percent more than women, for both heterosexual and nonheterosexual employees. This gender pay gap is robust across a range of specifications. No statistically significant pay gap was found between heterosexual and LGB employees, although decomposition analysis suggested that offsetting effects for those who disclose their sexual orientation were masking pay gaps within this group relative to comparable heterosexuals. Individuals who have disclosed LGB orientation to their work colleagues received some 6 percent higher wages than heterosexuals, whereas those who have not disclosed face a similar-sized wage penalty. This is true for both genders. Among LGB employees, disclosure was associated with 13 percent more pay, with some three quarters of this gap related to unexplained differences in returns to observable characteristics. The results suggest that LGB employees who have disclosed their sexual identity in their workplace received more favorable pay rewards given their endowments relative to their closeted counterparts (sometimes referred to as positive discrimination).

The finding was particularly strong among male GB workers; those who have disclosed their sexual identity earn some 18 lpp more than those who have not disclosed, with some 7 lpp from those who disclose having more of the characteristics associated with higher pay, and some 11 lpp from receiving a higher return (favorable rewards) on their characteristics. These results suggest that there is substantial heterogeneity within LGB employees in the NHS—between those who have chosen to disclose their sexual identity and those who have chosen to remain closeted in the workplace.

The EES-NHS survey used in this study considered only LGB and heterosexual employees in the health sector; in particular, the English NHS. The resulting sample sizes are reasonable; however, they are not large compared to potential response rates. Both of these factors limited the ability to legitimately extrapolate the findings to a broader social context and suggest a need for further studies. Nevertheless, this study revealed heterogeneity within LGB workers and found that disclosure was related to more favorable pay treatment for LGB employees within the NHS. Studies exploring the determinants of the disclosure decision in the workplace will help to explain these findings more fully in the future.

Acknowledgments

The authors thank Yvonne Berks, Melanie Jones, Peter Smith, members of the Project Research Advisory Board including representatives from NHS employers; Stonewall; and Employers Networks for Equality and Inclusion. Funding was provided by the Economic and Social Research Council (ESRC) [grant number ES/N019334/1].

Notes

1. https://www.gov.uk/government/organisations/nhs-pay-review-body
2. https://digital.nhs.uk/data-and-information/publications/statistical/nhs-workforce-statistics/september-2018
3. https://government-equalities-office.shinyapps.io/lgbt-survey-2017/
4. https://www.gov.uk/government/organisations/nhs-pay-review-body
5. Jones and Ezgi (2019; page 40, Table 3) found a similar raw gender pay gap of 5.4 percent for the NHSPRB employees using pooled Labour Force Survey data between 2016 and 2018.
6. There were 17 self-identifying transgender individuals in the sample. When further asked "Which of the following best described how you think of yourself?", their responses were as follows: seven heterosexual, three gay/lesbian, five bisexual, one don't know, and one other. Removing the transgender individuals from the sample did not change the findings in any substantial or statistically significant manner. Results are available from the authors upon request.
7. Complete results are presented in appendix; results for the full model are presented in column 7 of Table A2, with selected results for pay gaps and disclosure presented in panel a of Table 2. Comparable results for males (GB and heterosexuals) are

reported in Table A3 (column 7) and panel b of Table 2, with comparable results for females in Table A4 (column 7) and panel c of Table 2.

8. There is a pay penalty associated with being disabled. That penalty was of a similar size for males and females but not statistically significant for males. The measurement and interpretation of a disabled pay penalty is a complex issue that will be considered in future work by the authors.

9. Limited evidence of pay differences for bisexual employees was found, either in aggregate or gender-specific analyses (results are available upon request).

10. Comparing results from the full model applied to all employees (column 7 of Table A2) or the full model applied to the sample of employees who are living with a partner (column 8 of Table A2) revealed no significant difference in any of the findings. The pay premiums associated with LGB status and disclosure were not as well defined, as would be expected with a smaller sample size, but they showed little qualitative difference. This was true also for males (comparing columns 7 and 8 in Table A3) and females (comparing columns 7 and 8 in Table A4). There was also no significant difference in results for those who are formally married or living together (results available upon request).

11. $\hat{\beta}_b$ and $\hat{\beta}_a$ are reported in the Table A4, columns 2 and 3, respectively; and $\overline{X}_a$ and $\overline{X}_b$ are reported in Table A1, columns 2 and 3 respectively.

Appendix

Table A1. Definitions and means (standard deviations) of variables

	Means (standard deviations)					Definitions
	Pooled	Male	Female	LGB	Heterosexual	
	(1)	(2)	(3)	(4)	(5)	
salary	16.62	***17.36***	***16.42***	16.83	16.59	Average hourly pay in GBP (full time equivalent)
	(6.53)	(7.27)	(6.30)	(6.80)	(6.49)	
natural log salary	2.742	***2.777***	***2.733***	2.754	2.741	
	(0.36)	(0.39)	(0.35)	(0.36)	(0.36)	
Sexual Identity						
LGB	0.12	***0.28***	***0.08***			Respondent LGB
disclose	0.06	***0.17***	***0.03***	0.51		Open about sexual identity in current job
Demographics						
male	0.21			***0.48***	***0.17***	Respondent is male
age	46.21	***45.03***	***46.53***	***41.46***	***46.88***	Age of respondent
	(11.43)	(11.79)	(11.31)	(11.32)	(11.28)	
ethnic minority	0.11	***0.14***	***0.10***	**0.09**	**0.12**	Ethnic group not white (Mixed; Asian; Black/Black British; Arab or Other)
married	0.51	0.49	0.51	***0.29***	***0.54***	Married
live in couples	0.69	0.70	0.68	***0.57***	***0.70***	Living together
dependent children	0.32	0.30	0.32	***0.14***	***0.34***	Has dependent children
disability	0.36	0.37	0.35	***0.45***	***0.34***	Long-standing illness, health problem or disability
caregiver responsibilities	0.26	***0.19***	***0.28***	**0.23**	**0.27**	Look after or give support to family/friend due to health or old age
foreign	0.12	**0.14**	**0.12**	***0.09***	***0.13***	Born outside UK
Qualifications						
min qual	0.01	0.01	0.01		0.01	Below O level or no recognized qualifications
GCSE D-G	0.05	0.04	0.05	***0.02***	***0.05***	O level or GCSE grades D-G
GCSE A-C	0.08	**0.07**	**0.09**	***0.05***	***0.09***	O level or GCSE grades A-C
trade	0.004	***0.01***	***0.002***	0.002	0.004	Trade apprenticeships
A levels	0.09	0.10	0.09	0.10	0.09	A levels and others
HE and TQ	0.16	0.17	0.16	0.15	0.16	Diploma in higher education, teaching qualifications and others
first degree	0.30	0.31	0.30	0.32	0.30	First degree and PGCE
higher degree	0.28	0.27	0.28	***0.32***	***0.27***	Higher degree or postgraduate
potential experience	17.96	***16.02***	***18.48***	***15.12***	***18.36***	Years of potential experience
	(11.66)	(11.11)	(11.75)	(10.57)	(11.75)	

	Means (standard deviations)					Definitions
	Pooled	Male	Female	LGB	Heterosexual	
	(1)	(2)	(3)	(4)	(5)	
Occupation						
allied	0.19	0.19	0.19	0.21	0.19	Allied health professional, healthcare scientist, scientific and technical
ambulance	0.009	***0.03***	***0.003***	***0.04***	***0.004***	Ambulance (operational)
public health	0.01	**0.01**	**0.009**	0.01	0.01	Public health/health improvement
commissioning manager	0.01	***0.02***	***0.01***	0.02	0.01	Commissioning manager/support staff
nurses	0.24	***0.14***	***0.27***	**0.21**	**0.25**	Registered nurse and midwives
nursing auxiliary	0.05	0.05	0.05	0.06	0.05	Nursing auxiliary, nursing assistant, health care assistants
social care	0.006	0.007	0.006	0.004	0.007	Social care
wider	0.24	***0.21***	***0.25***	***0.19***	***0.25***	Wider healthcare team
general management	0.09	***0.14***	***0.08***	0.11	0.09	General management
other	0.10	***0.15***	***0.09***	0.10	0.10	Other
health professional	0.44	***0.35***	***0.46***	0.43	0.44	Respondent is health professional
Job Characteristics						
part time	0.24	***0.09***	***0.28***	***0.11***	***0.26***	Part-time work
job permanent	0.93	0.92	0.93	0.93	0.93	Has a permanent contract
trade union	0.57	***0.52***	***0.58***	0.56	0.57	Member of a trade union
tenure	6.87	***6.20***	***7.05***	***5.53***	***7.06***	Years in current job
Workplace Characteristics						
mentor	0.47	***0.42***	***0.48***	0.45	0.47	Has mentor/coach for work advice
happy training	0.47	0.45	0.47	0.48	0.47	Satisfied with opportunities to develop skills
friend	0.61	***0.49***	***0.64***	***0.56***	***0.61***	At least one close friend in workplace
cooperative	0.39	0.41	0.39	0.43	0.39	Feel workplace is cooperative
bully witnessed	1.84	1.86	1.84	1.83	1.84	Has witnessed bullying at work
discrimination	1.95	1.95	1.95	1.97	1.95	Has been subject to discrimination at work
responsive hours	0.46	***0.39***	***0.48***	0.46	0.46	Using at least one of the responsive work hours (flexi-time, reduced hours, same hours fewer days and paid leave to care)
pressure	0.55	0.55	0.56	0.55	0.56	Job makes feel pressure always and often
coworker support	0.77	0.75	0.78	0.80	0.77	Has supportive colleagues
work-life balance	0.59	0.59	0.59	0.61	0.59	Maintains work-life balance
supervisor support	0.61	0.59	0.62	0.60	0.61	Supervisor responds to suggestions

	Means (standard deviations)					Definitions
	Pooled	**Male**	**Female**	**LGB**	**Heterosexual**	
	(1)	(2)	(3)	(4)	(5)	
NHS England Region						
North of England	0.23	0.23	0.23	0.23	0.23	
Midlands and East of England	0.33	***0.28***	***0.34***	***0.25***	***0.34***	
London	0.15	***0.17***	***0.14***	***0.24***	***0.14***	
South West	0.11	0.12	0.11	***0.07***	***0.12***	
South East	0.15	***0.18***	***0.15***	***0.19***	***0.15***	
Trust Type						
Acute Specialist Trusts	0.02	***0.03***	***0.01***	***0.05***	***0.01***	
Acute Trusts	0.50	0.51	0.50	***0.38***	***0.51***	
Ambulance Trusts	0.01	***0.03***	***0.008***	***0.04***	***0.008***	
Combined Acute and Community Trusts	0.12	0.12	0.12	0.11	0.12	
Combined Mental Health / Learning Disability and Community Trusts	0.08	**0.07**	**0.09**	***0.11***	***0.08***	
Community Trusts	0.10	***0.05***	***0.11***	**0.08**	**0.10**	
Mental Health / Learning Disability Trusts	0.14	0.15	0.13	***0.20***	***0.13***	
Observations	3,556	753	2,803	440	3,116	

Mean pair differences: males (2) vs. females (3); LGB (3) vs. heterosexual (4)

bold p < 0.10, ***bold and italic p < 0.05***

Table A2. The determinants of log earnings (OLS estimates), full sample

	(1)	(2)	(3)	(4)	(5)	(6)	(7)	(8)	(9)
Dependent variable is ln(salary)	**Male**	**LGB**	**Base**	**+HC**	**+Demog**	**+Occup**	**+Job/Work**	**Full**	**Coupled**
male	0.0440***	0.0440***	0.0396***	0.0660***	0.0624***	0.0500***	0.0423***	0.0388***	0.0509***
	(0.0139)	(0.0140)	(0.0137)	(0.0121)	(0.0118)	(0.0106)	(0.0111)	(0.0109)	(0.0134)
LGB		–0.0001							
		(0.0193)							
no disclose & LGB			–0.0612**	–0.0558***	–0.0323*	–0.0352**	–0.0411**	–0.0493***	–0.0495**
			(0.0238)	(0.0175)	(0.0171)	(0.0157)	(0.0159)	(0.0165)	(0.0232)
disclose & LGB			0.0614**	0.0330*	0.0572***	0.0577***	0.0557***	0.0435**	0.0364
			(0.0263)	(0.0193)	(0.0187)	(0.0178)	(0.0180)	(0.0172)	(0.0253)
Qualifications (omitted group: min qual)									
O level				0.0738	0.0865**	0.0087	0.0030	0.0001	0.0501
				(0.0488)	(0.0437)	(0.0506)	(0.0507)	(0.0517)	(0.0612)
GCSE				0.1440***	0.1505***	0.0424	0.0362	0.0275	0.0892
				(0.0537)	(0.0488)	(0.0522)	(0.0521)	(0.0531)	(0.0679)
trade				0.2148**	0.2165**	0.1102	0.0920	0.0938	0.1905*
				(0.0962)	(0.0962)	(0.0915)	(0.0787)	(0.0794)	(0.1004)
A levels				0.1990***	0.2104***	0.0987**	0.0893*	0.0854*	0.1327**
				(0.0451)	(0.0413)	(0.0473)	(0.0476)	(0.0484)	(0.0594)
HE and TQ				0.3615***	0.3675***	0.2017***	0.1881***	0.1847***	0.2437***
				(0.0483)	(0.0439)	(0.0475)	(0.0468)	(0.0481)	(0.0575)
first degree				0.5134***	0.5137***	0.3128***	0.2859***	0.2807***	0.3476***
				(0.0485)	(0.0441)	(0.0499)	(0.0478)	(0.0491)	(0.0604)
higher degree				0.6842***	0.6862***	0.4630***	0.4288***	0.4176***	0.4826***
				(0.0496)	(0.0449)	(0.0510)	(0.0488)	(0.0508)	(0.0595)
potential experience				0.0238***	0.0220***	0.0174***	0.0163***	0.0156***	0.0166***
				(0.0018)	(0.0018)	(0.0017)	(0.0018)	(0.0017)	(0.0022)
potential experience^2				–0.0003***	–0.0003***	–0.0002***	–0.0002***	–0.0002***	–0.0002***
				(0.0000)	(0.0000)	(0.0000)	(0.0000)	(0.0000)	
age					0.0003	0.0005	0.0016***	0.0016***	0.0015**
					(0.0005)	(0.0004)	(0.0004)	(0.0004)	(0.0006)
ethnic minority					0.0280	0.0180	0.0139	–0.0146	–0.0230
					(0.0177)	(0.0151)	(0.0152)	(0.0144)	(0.0192)

	(1)	(2)	(3)	(4)	(5)	(6)	(7)	(8)	(9)
Dependent variable is ln(salary)	**Male**	**LGB**	**Base**	**+HC**	**+Demog**	**+Occup**	**+Job/Work**	**Full**	**Coupled**
live in couples					0.0494***	0.0424***	0.0415***	0.0474***	
					(0.0094)	(0.0085)	(0.0081)	(0.0083)	
dependent children					0.0468***	0.0367***	0.0453***	0.0470***	0.0576***
					(0.0103)	(0.0091)	(0.0092)	(0.0091)	(0.0105)
disability					–0.0464***	–0.0348***	–0.0299***	–0.0311***	–0.0313***
					(0.0113)	(0.0091)	(0.0089)	(0.0087)	(0.0108)
caregiving responsibilities					0.0016	0.0061	0.0027	0.0031	0.0045
					(0.0091)	(0.0090)	(0.0085)	(0.0081)	(0.0112)
foreign					–0.0203	–0.0165	–0.0163	–0.0295**	–0.0353**
					(0.0168)	(0.0150)	(0.0146)	(0.0136)	(0.0166)
Occupational Group (omitted group: registered nurse and midwives)									
allied						0.0238*	0.0300**	0.0324**	0.0435**
						(0.0127)	(0.0127)	(0.0125)	(0.0178)
ambulance						0.0181	0.0278	0.0438	0.0184
						(0.0479)	(0.0472)	(0.0632)	(0.0634)
public health						0.0178	0.0229	0.0230	0.0082
						(0.0375)	(0.0378)	(0.0382)	(0.0487)
commissioning manager						0.1526***	0.1372***	0.1344***	0.1692***
						(0.0347)	(0.0322)	(0.0318)	(0.0407)
nursing auxiliary						–0.1401***	–0.1362***	–0.1343***	–0.1350***
						(0.0226)	(0.0214)	(0.0202)	(0.0288)
social care						0.1690***	0.1526***	0.1640***	0.1739***
						(0.0385)	(0.0399)	(0.0414)	(0.0564)
wider						0.0762***	0.0640***	0.0674***	0.0762***
						(0.0208)	(0.0196)	(0.0193)	(0.0269)
general management						0.4170***	0.3747***	0.3710***	0.3737***
						(0.0245)	(0.0239)	(0.0223)	(0.0315)
other						0.0565***	0.0439**	0.0461**	0.0611**
						(0.0201)	(0.0203)	(0.0197)	(0.0269)
health professional						0.2265***	0.2316***	0.2288***	0.2271***
						(0.0179)	(0.0174)	(0.0167)	(0.0225)
part time							–0.0874***	–0.0838***	–0.0742***
							(0.0111)	(0.0103)	(0.0122)

	(1)	(2)	(3)	(4)	(5)	(6)	(7)	(8)	(9)
Dependent variable is ln(salary)	**Male**	**LGB**	**Base**	**+HC**	**+Demog**	**+Occup**	**+Job/Work**	**Full**	**Coupled**
job permanent							–0.0091	–0.0057	–0.0055
							(0.0173)	(0.0162)	(0.0184)
trade union							–0.0319***	–0.0286***	–0.0367***
							(0.0098)	(0.0094)	(0.0130)
mentor							–0.0516***	–0.0464***	–0.0488***
							(0.0074)	(0.0078)	(0.0102)
happy training							0.0729***	0.0689***	0.0739***
							(0.0100)	(0.0102)	(0.0130)
friend							0.0132	0.0102	0.0020
							(0.0082)	(0.0083)	(0.0091)
responsive hours							0.0407***	0.0433***	0.0402***
							(0.0096)	(0.0087)	(0.0138)
pressure							0.0310***	0.0343***	0.0382***
							(0.0085)	(0.0084)	(0.0108)
coworker support							0.0189*	0.0184*	0.0150
							(0.0098)	(0.0099)	(0.0133)
work–life balance							–0.0294***	–0.0278***	–0.0325***
							(0.0091)	(0.0088)	(0.0108)
supervisor support							0.0306***	0.0287***	0.0290**
							(0.0097)	(0.0097)	(0.0127)
cooperative								0.0088	0.0102
								(0.0081)	(0.0101)
NHS England Region (omitted group: North of England)									
Midlands and East of England								0.0252*	0.0296
								(0.0152)	(0.0180)
London								0.1224***	0.1403***
								(0.0169)	(0.0216)
South West								0.0065	0.0058
								(0.0193)	(0.0244)
South East								0.0378**	0.0478**
								(0.0161)	(0.0201)

	(1)	(2)	(3)	(4)	(5)	(6)	(7)	(8)	(9)
Dependent variable is ln(salary)	**Male**	**LGB**	**Base**	**+HC**	**+Demog**	**+Occup**	**+Job/Work**	**Full**	**Coupled**
Trust Type (omitted group: Acute Trusts)									
Acute Specialist Trusts								0.0216	0.0144
								(0.0306)	(0.0353)
Ambulance Trusts								–0.0051	0.0419
								(0.0698)	(0.0608)
Combined Acute and Community Trusts								0.0000	0.0092
								(0.0165)	(0.0205)
Combined Mental Health / Learning Disability and Community Trusts								-0.0437**	–0.0321
								(0.0209)	(0.0239)
Community Trusts								–0.0415**	–0.0485***
								(0.0179)	(0.0164)
Mental Health / Learning Disability Trusts								–0.0084	–0.0002
								(0.0132)	(0.0165)
Constant	2.7334***	2.7334***	2.7342***	2.0094***	1.9695***	2.0368***	1.9993***	1.9776***	1.9485***
	(0.0159)	(0.0165)	(0.0166)	(0.0514)	(0.0511)	(0.0569)	(0.0554)	(0.0561)	(0.0784)
Observations	3556	3556	3556	3556	3556	3556	3556	3556	2443
R^2	0.002	0.002	0.006	0.457	0.470	0.580	0.611	0.6241	0.6205
Adj. R^2	0.002	0.001	0.005	0.455	0.467	0.577	0.607	0.6186	0.6126

Standard errors are in parentheses (clustered at Trust level).

*$p < 0.10$, **$p < 0.05$, ***$p < 0.01$

Table A3. The determinants of log earnings (OLS estimates), male subsample

Dependent variable is ln(salary)	(1) LGB	(2) Base	(3) +HC	(4) +Demog	(5) +Occup	(6) +Job/Work	(7) Full	(8) Coupled
LGB	0.0258							
	(0.0281)							
no disclose & LGB		–0.0806**	–0.0666**	–0.0173	–0.0271	–0.0235	–0.0254	–0.0400
		(0.0360)	(0.0300)	(0.0300)	(0.0312)	(0.0295)	(0.0280)	(0.0390)
disclose & LGB		0.0953***	0.0370	0.0958***	0.0925***	0.0830***	0.0722***	0.0692
		(0.0361)	(0.0286)	(0.0299)	(0.0290)	(0.0294)	(0.0270)	(0.0423)
Qualifications (omitted group: min qual)								
O level			–0.0780	–0.0607	–0.1015	–0.1198	–0.1500	–0.2223**
			(0.1314)	(0.1323)	(0.1204)	(0.0910)	(0.0972)	(0.1078)
GCSE			–0.0053	0.0259	–0.0619	–0.1089	–0.1460	–0.1508
			(0.1276)	(0.1248)	(0.1220)	(0.0937)	(0.1059)	(0.1266)
trade			0.1218	0.1336	0.0400	0.0213	–0.0023	–0.0139
			(0.2183)	(0.2133)	(0.2056)	(0.1712)	(0.1751)	(0.1987)
A levels			0.0461	0.0844	0.0105	–0.0014	–0.0242	–0.0662
			(0.0994)	(0.0985)	(0.1064)	(0.0773)	(0.0871)	(0.1006)
HE and TQ			0.2514**	0.2780**	0.1460	0.1059	0.0726	0.0809
			(0.1202)	(0.1194)	(0.1214)	(0.0856)	(0.0962)	(0.0996)
first degree			0.3802***	0.3983***	0.2402*	0.2036**	0.1726*	0.1561
			(0.1275)	(0.1225)	(0.1261)	(0.0914)	(0.1008)	(0.1087)
higher degree			0.5323***	0.5543***	0.3539***	0.3098***	0.2745***	0.2624***
			(0.1144)	(0.1088)	(0.1179)	(0.0823)	(0.0948)	(0.1002)
potential experience			0.0265***	0.0212***	0.0202***	0.0178***	0.0170***	0.0190***
			(0.0037)	(0.0039)	(0.0036)	(0.0036)	(0.0034)	(0.0038)
potential experience squared			–0.0004***	–0.0003***	–0.0003***	–0.0003***	–0.0003***	–0.0003***
			(0.0001)	(0.0001)	(0.0001)	(0.0001)	(0.0001)	(0.0001)
age				0.0023*	0.0015	0.0027**	0.0031***	0.0034**
				(0.0013)	(0.0012)	(0.0012)	(0.0011)	(0.0013)
ethnic minority				0.0387	0.0193	0.0214	–0.0042	0.0112
				(0.0353)	(0.0322)	(0.0308)	(0.0333)	(0.0434)
live in couples				0.0889***	0.0832***	0.0685***	0.0681***	
				(0.0204)	(0.0190)	(0.0198)	(0.0181)	

Dependent variable is	(1)	(2)	(3)	(4)	(5)	(6)	(7)	(8)
ln(salary)	**LGB**	**Base**	**+HC**	**+Demog**	**+Occup**	**+Job/Work**	**Full**	**Coupled**
dependent children				0.0887***	0.0691***	0.0591***	0.0722***	0.0698**
				(0.0251)	(0.0227)	(0.0220)	(0.0216)	(0.0270)
disability				–0.0397	–0.0195	–0.0217	–0.0228	–0.0119
				(0.0253)	(0.0214)	(0.0205)	(0.0210)	(0.0259)
caregiving responsibilities				0.0050	0.0097	0.0157	0.0123	0.0199
				(0.0262)	(0.0209)	(0.0208)	(0.0190)	(0.0237)
foreign				–0.0635**	–0.0326	–0.0414	–0.0517*	–0.0598
				(0.0319)	(0.0299)	(0.0302)	(0.0311)	(0.0393)
Occupational Group (omitted group: registered nurse and midwives)								
allied					0.0524*	0.0553*	0.0546*	0.0794*
					(0.0298)	(0.0308)	(0.0307)	(0.0409)
ambulance					0.0569	0.0667	0.2914***	0.3320***
					(0.0537)	(0.0505)	(0.0510)	(0.0663)
public health					–0.0063	0.0059	–0.0243	–0.0925
					(0.0783)	(0.0792)	(0.0950)	(0.0927)
commissioning manager					0.1168*	0.1100**	0.1210**	0.1495**
					(0.0611)	(0.0535)	(0.0532)	(0.0704)
nursing auxiliary					–0.1880***	–0.2005***	–0.1876***	–0.1547***
					(0.0482)	(0.0469)	(0.0470)	(0.0536)
social care					0.4024***	0.3301**	0.3991***	0.4246***
					(0.1471)	(0.1544)	(0.1405)	(0.1493)
wider					0.1022**	0.0915*	0.0972**	0.1669**
					(0.0484)	(0.0468)	(0.0439)	(0.0643)
general management					0.3639***	0.3268***	0.3400***	0.3446***
					(0.0615)	(0.0587)	(0.0556)	(0.0691)
other					0.0477	0.0369	0.0316	0.0737
					(0.0502)	(0.0493)	(0.0472)	(0.0572)
health professional					0.1842***	0.1775***	0.1868***	0.1795***
					(0.0426)	(0.0428)	(0.0411)	(0.0547)
part time						–0.0858**	–0.0792**	–0.0051
						(0.0336)	(0.0318)	(0.0398)
job permanent						0.0426	0.0503*	0.1327***
						(0.0302)	(0.0288)	(0.0489)

Dependent variable is ln(salary)	(1) LGB	(2) Base	(3) +HC	(4) +Demog	(5) +Occup	(6) +Job/Work	(7) Full	(8) Coupled
trade union						−0.0183	−0.0139	−0.0273
						(0.0237)	(0.0242)	(0.0290)
mentor						−0.0495**	−0.0425*	−0.0375
						(0.0238)	(0.0236)	(0.0304)
happy training						0.0680***	0.0677***	0.0707***
						(0.0216)	(0.0211)	(0.0257)
friend						0.0544***	0.0453**	0.0324
						(0.0196)	(0.0200)	(0.0231)
responsive hours						0.0259	0.0334	0.0166
						(0.0233)	(0.0230)	(0.0323)
pressure						0.0644***	0.0646***	0.0848***
						(0.0214)	(0.0206)	(0.0282)
coworker support						0.0405	0.0366	0.0321
						(0.0254)	(0.0253)	(0.0358)
work–life balance						−0.0319	−0.0199	−0.0151
						(0.0239)	(0.0238)	(0.0264)
supervisor support						0.0539**	0.0555**	0.0792***
						(0.0245)	(0.0243)	(0.0296)
cooperative							−0.0009	−0.0017
							(0.0221)	(0.0279)
NHS England Region (omitted group: North of England)								
Midlands and East of England							0.0470	0.0400
							(0.0304)	(0.0374)
London							0.1758***	0.1847***
							(0.0325)	(0.0427)
South West							0.0497	0.0596
							(0.0376)	(0.0434)
South East							0.0524	0.0660*
							(0.0317)	(0.0394)
Trust Type (omitted group: Acute Trusts)								
Acute Specialist Trusts							−0.0441	−0.0496
							(0.0511)	(0.0730)
Ambulance Trusts							−0.2429***	−0.2298***
							(0.0471)	(0.0715)

Dependent variable is ln(salary)	**(1) LGB**	**(2) Base**	**(3) +HC**	**(4) +Demog**	**(5) +Occup**	**(6) +Job/Work**	**(7) Full**	**(8) Coupled**
Combined Acute and Community Trusts							0.0079	0.0317
							(0.0334)	(0.0411)
Combined Mental Health / Learning Disability and Community Trusts							–0.1338***	–0.1436***
							(0.0341)	(0.0446)
Community Trusts							0.0043	–0.0010
							(0.0293)	(0.0374)
Mental Health / Learning Disability Trusts							–0.0267	0.0071
							(0.0258)	(0.0345)
Constant	2.7702***	2.7702***	2.1999***	2.0399***	2.0798***	1.9633***	1.9228***	1.8400***
	(0.0205)	(0.0205)	(0.1175)	(0.1140)	(0.1178)	(0.0939)	(0.0992)	(0.1144)
Observations	753	753	753	753	753	753	753	524
R^2	0.0009	0.0148	0.4031	0.4364	0.5386	0.5732	0.6018	0.5786
Adj. R^2	–0.0004	0.0122	0.3943	0.4226	0.5208	0.5499	0.5734	0.5350

Standard errors are in parentheses (clustered at Trust level).

*p < 0.10, **p < 0.05, ***p < 0.01

Table A4. The determinants of log earnings (OLS estimates), female subsample

	(1)	(2)	(3)	(4)	(5)	(6)	(7)	(8)
	LGB	**Base**	**+HC**	**+Demog**	**+Occup**	**+Job/Work**	**Full**	**Coupled**
LGB	–0.0187							
	(0.0255)							
no disclose & LGB		–0.0477	–0.0473**	–0.0291	–0.0282	–0.0395**	–0.0500**	–0.0439
		(0.0291)	(0.0234)	(0.0221)	(0.0184)	(0.0178)	(0.0194)	(0.0273)
disclose & LGB		0.0210	0.0274	0.0406	0.0561**	0.0489**	0.0394*	0.0323
		(0.0377)	(0.0261)	(0.0250)	(0.0238)	(0.0241)	(0.0232)	(0.0312)
Qualifications (omitted group: min quals)								
O level			0.1116***	0.1167***	0.0347	0.0358	0.0335	0.1201***
			(0.0313)	(0.0309)	(0.0404)	(0.0390)	(0.0399)	(0.0353)
GCSE			0.1812***	0.1767***	0.0657	0.0677	0.0619	0.1504***
			(0.0350)	(0.0341)	(0.0414)	(0.0414)	(0.0415)	(0.0388)
trade			0.2039***	0.1837***	0.1020*	0.0786	0.0747	0.1784***
			(0.0701)	(0.0599)	(0.0578)	(0.0543)	(0.0501)	(0.0523)
A levels			0.2382***	0.2361***	0.1192***	0.1136***	0.1105***	0.1890***
			(0.0329)	(0.0338)	(0.0403)	(0.0400)	(0.0398)	(0.0334)
HE and TQ			0.3879***	0.3808***	0.2102***	0.2041***	0.2014***	0.2810***
			(0.0337)	(0.0349)	(0.0416)	(0.0414)	(0.0422)	(0.0361)
first degree			0.5465***	0.5340***	0.3249***	0.3032***	0.2975***	0.3911***
			(0.0300)	(0.0308)	(0.0412)	(0.0396)	(0.0400)	(0.0344)
higher degree			0.7224***	0.7110***	0.4863***	0.4557***	0.4451***	0.5386***
			(0.0322)	(0.0322)	(0.0422)	(0.0411)	(0.0424)	(0.0359)
potential experience			0.0232***	0.0225***	0.0169***	0.0162***	0.0158***	0.0165***
			(0.0018)	(0.0018)	(0.0016)	(0.0018)	(0.0017)	(0.0022)
potential experience squared			–0.0003***	–0.0003***	–0.0002***	–0.0002***	–0.0002***	–0.0002***
			(0.0000)	(0.0000)	(0.0000)	(0.0000)	(0.0000)	(0.0000)
age				–0.0006	0.0001	0.0012**	0.0012**	0.0012*
				(0.0007)	(0.0005)	(0.0005)	(0.0005)	(0.0007)
ethnic minority				0.0236	0.0145	0.0077	–0.0238	–0.0324
				(0.0226)	(0.0181)	(0.0175)	(0.0165)	(0.0230)
live in couples				0.0383***	0.0322***	0.0341***	0.0389***	
				(0.0105)	(0.0096)	(0.0090)	(0.0092)	

	(1)	(2)	(3)	(4)	(5)	(6)	(7)	(8)
	LGB	**Base**	**+HC**	**+Demog**	**+Occup**	**+Job/Work**	**Full**	**Coupled**
dependent children				0.0368***	0.0306***	0.0430***	0.0435***	0.0611***
				(0.0114)	(0.0100)	(0.0100)	(0.0097)	(0.0108)
disability				–0.0485***	–0.0372***	–0.0306***	–0.0323***	–0.0338***
				(0.0118)	(0.0099)	(0.0097)	(0.0094)	(0.0115)
caregiving responsibilities				0.0032	0.0066	0.0010	0.0010	0.0035
				(0.0104)	(0.0095)	(0.0086)	(0.0083)	(0.0118)
foreign				–0.0063	–0.0101	–0.0097	–0.0218	–0.0267
				(0.0182)	(0.0160)	(0.0156)	(0.0151)	(0.0185)
Occupational Group (omitted group: registered nurse and midwives)								
allied					0.0227	0.0275**	0.0291**	0.0377**
					(0.0144)	(0.0136)	(0.0136)	(0.0189)
ambulance					–0.0575	–0.0665	–0.1601***	–0.1768***
					(0.0607)	(0.0584)	(0.0594)	(0.0670)
public health					0.0280	0.0326	0.0426	0.0400
					(0.0488)	(0.0459)	(0.0430)	(0.0523)
commissioning manager					0.1823***	0.1617***	0.1553***	0.1994***
					(0.0364)	(0.0355)	(0.0352)	(0.0466)
nursing auxiliary					–0.1145***	–0.1087***	–0.1090***	–0.1073***
					(0.0263)	(0.0249)	(0.0241)	(0.0355)
social care					0.1026**	0.0997**	0.0972**	0.0794**
					(0.0440)	(0.0432)	(0.0478)	(0.0402)
wider					0.0881***	0.0733***	0.0742***	0.0762***
					(0.0224)	(0.0206)	(0.0207)	(0.0287)
general management					0.4538***	0.4082***	0.3972***	0.4097***
					(0.0252)	(0.0239)	(0.0227)	(0.0326)
other					0.0729***	0.0577**	0.0584**	0.0695**
					(0.0232)	(0.0224)	(0.0225)	(0.0323)
health professional					0.2506***	0.2602***	0.2556***	0.2615***
					(0.0204)	(0.0198)	(0.0197)	(0.0268)
part time						–0.0856***	–0.0815***	–0.0776***
						(0.0110)	(0.0106)	(0.0122)
job permanent						–0.0227	–0.0204	–0.0183
						(0.0201)	(0.0191)	(0.0204)

	(1)	(2)	(3)	(4)	(5)	(6)	(7)	(8)
	LGB	Base	+HC	+Demog	+Occup	+Job/Work	Full	Coupled
trade union						–0.0379***	–0.0350***	–0.0430***
						(0.0104)	(0.0099)	(0.0136)
mentor						–0.0531***	–0.0485***	–0.0515***
						(0.0079)	(0.0078)	(0.0123)
happy training						0.0728***	0.0677***	0.0718***
						(0.0105)	(0.0108)	(0.0126)
friend						0.0036	0.0014	–0.0032
						(0.0091)	(0.0092)	(0.0110)
responsive hours						0.0440***	0.0452***	0.0425***
						(0.0093)	(0.0087)	(0.0111)
pressure						0.0222**	0.0255***	0.0268**
						(0.0094)	(0.0095)	(0.0112)
coworker support						0.0122	0.0116	0.0047
						(0.0128)	(0.0130)	(0.0167)
work–life balance						–0.0288***	–0.0274***	–0.0321**
						(0.0092)	(0.0089)	(0.0125)
supervisor support						0.0259**	0.0242**	0.0197
						(0.0107)	(0.0100)	(0.0130)
cooperative							0.0110	0.0144
							(0.0088)	(0.0107)
NHS England Region (omitted group: North of England)								
Midlands and East of England							0.0210	0.0248
							(0.0165)	(0.0194)
London							0.1067***	0.1219***
							(0.0174)	(0.0206)
South West							–0.0031	–0.0110
							(0.0197)	(0.0231)
South East							0.0362**	0.0421**
							(0.0164)	(0.0206)
Trust Type (omitted group: Acute Trusts)								
Acute Specialist Trusts							0.0498	0.0423
							(0.0341)	(0.0361)
Ambulance Trusts							0.1120*	0.1331*
							(0.0655)	(0.0705)

	(1)	(2)	(3)	(4)	(5)	(6)	(7)	(8)
	LGB	**Base**	**+HC**	**+Demog**	**+Occup**	**+Job/Work**	**Full**	**Coupled**
Combined Acute and Community Trusts							0.0034	0.0102
							(0.0168)	(0.0207)
Combined Mental Health / Learning Disability and Community Trusts							–0.0198	0.0011
							(0.0238)	(0.0257)
Community Trusts							–0.0468**	–0.0523***
							(0.0180)	(0.0187)
Mental Health / Learning Disability Trusts							–0.0021	0.0048
							(0.0157)	(0.0190)
Constant	2.7349***	2.7349***	1.9777***	1.9855***	2.0290***	2.0128***	1.9988***	1.9336***
	(0.0169)	(0.0169)	(0.0335)	(0.0476)	(0.0532)	(0.0564)	(0.0579)	(0.0658)
Observations	2803	2803	2803	2803	2803	2803	2803	1919
R^2	0.0002	0.0010	0.4739	0.4851	0.6005	0.6319	0.6431	0.6508
Adj. R^2	–0.0001	0.0002	0.4718	0.4818	0.5965	0.6267	0.6367	0.6416

Standard errors are in parentheses (clustered at Trust level).

*$p < 0.10$, **$p < 0.05$, ***$p < 0.01$

References

Aksoy, C.G., Carpenter, C.S., and Frank. J. 2018. "Sexual orientation and earnings: New evidence from the United Kingdom." *International Labor Review*, Vol. 71, no. 1, pp. 242–272.

Aksoy, C.G., Carpenter, C.S., and Huffman, M.L. 2019. "Gay glass ceilings: Sexual orientation and workplace authority in the UK." *Journal of Economic Behavior and Organization*, Vol. 159, pp. 167–180.

Arabsheibani, R., Marin, A., and Wadsworth, J. 2005. "Gay pay in the UK." *Economica*, Vol. 72, no. 286, pp. 333–347.

Arrow, K. 1973. "The theory of discrimination." In *Discrimination in the Labor Markets*. Princeton, NJ: Princeton University Press, pp. 3–33.

Badgett, M.V.L. 1995. "The wage effects of sexual orientation discrimination." *Industrial and Labor Relations Review*, Vol. 48, no. 4, pp. 726–739.

Badgett, M.V.L. 1996. "Employment and sexual orientation: Disclosure and discrimination in the workplace." In A.L. Ellis and E.D.B. Riggle (eds.), *Sexual Identity on the Job: Issues and Services*. Harrington Park Press/Haworth Press, pp. 29–52.

Badgett, M.V.L. Carpenter, C.S., and Sasone, D. 2021. "LGBTQ economics." *Journal of Economic Perspectives*, Vol. 35, no. 2, pp. 141–171.

Becker, G. 1957. *The Economics of Discrimination*. Chicago: University of Chicago Press.

Becker, G. 1975. *Human Capital*, 2nd edition. Chicago: University of Chicago Press.

Becker, G. 1985. "Human capital, effort, and the sexual division of labor." *Journal of Labor Economics*, Vol. 3, no. 1, pp. S33–S38.

Blau, F., and Kahn. L. 2017. "The gender wage gap: Extent, trends, and sources." *Journal of Economic Literature*, Vol. 55, no. 3, pp. 789–865.

Bridges, S., and Mann, S. 2019. "Sexual orientation, legal partnerships and wages in Britain." *Work, Employment and Society*, Vol. 33, no. 6, pp. 1020–1038.

Bryson, A. 2017. "Pay equity after the Equality Act 2010: Does sexual orientation still matter?" *Work, Employment and Society*, Vol. 31, no. 3, pp. 484–500.

Einarsdóttir, A., Mumford, K., Birks, Y., Aguirre, E., Lockyer, B., and Sayli, M. 2020. "LGBT employee networks within the NHS: Technical report and data addendum." York: University of York. https://www.researchgate.net/publication/344670897

Fortin, N., Lemieux, T., and Firpo, S. 2011. "Decomposition methods in economics." In O. Ashenfelter and D. Card (eds.), *Handbook of Labor Economics* Volume 4(A). Amsterdam: Elsevier North Holland, pp. 1, 102.

Hudson-Sharp, N., and Metcalf, H. 2016. "Inequality among lesbian, gay bisexual and transgender groups in the UK: A review of evidence." Report. London: National Institute of Economic and Social Research.

Jepsen, C., and Jepsen, L. 2020. "Convergence over time or not? U.S. wages by sexual orientation, 2001–2018." IZA Discussion Paper 13495.

Jann, B. 2008. "The Blinder-Oaxaca decomposition for linear regression models." *The Stata Journal*, Vol. 8, no. 4, pp. 453–479.

Jones, M., and Ezgi, K. 2019. "Understanding the gender pay gap within the UK public sector." Research Report for Office of Manpower Economics (OME). Open Call for Research on Public Sector Pay and Workforces 2018.

Klawitter, M. 2015. "Meta-analysis of the effects of sexual orientation on earnings." *Industrial Relations: A Journal of Economy and Society*, Vol. 54, no. 1, pp. 4–32.

Mincer, J. 1974. *Schooling, Experience and Earnings*. New York: National Bureau of Economic Research.

Neumark, D. 1988. "Employers' discriminatory behavior and the estimation of wage discrimination." *Journal of Human Resources*, Vol. 23, no. 3, pp. 279–295.

Oaxaca, R. 1973. "Male–female wage differentials in urban labor markets." *International Economic Review*, Vol. 14, no. 3, pp. 693–709.

Oaxaca, R., and Ransom, M.R. 1994. "On discrimination and the decomposition of wage differentials." *Journal of Econometrics*, Vol. 61, pp. 5–21.

Phelps, E.S. 1972. "The statistical theory of racism and sexism." *The American Economic Review*, Vol. 62, pp. 659–661.

Plug, E., and Berkhout, P. 2008. "Sexual orientation, disclosure and earnings." IZA Discussion Paper No. 3290.

Spence, M. 1973. "Job market signalling." *Quarterly Journal of Economics*, Vol. 87, no. 3, pp. 355–374.

Wang, J., Gunderson, M., and Wicks, D. 2018. "The earnings effect of sexual orientation: British evidence from worker–firm matched data." *British Journal of Industrial Relations*, Vol. 56, pp. 744–769.

Wax, A., Coletti, K.K., and Ogaz, J.W. 2018. "The benefit of full disclosure: A meta-analysis of the implications of coming out at work." *Organizational Psychology Review*, Vol. 8, no. 1, pp. 3–30.

XI. LERA Annual Reports

LERA Executive Board Meeting Minutes
Thursday, February 10, 2022, 12 noon Eastern Time
Via Videoconference

Call to Order—The meeting was called to order at 12:01 p.m. by Wilma Liebman, President. Present at the meeting were board members Daniel Altchek, Peter Berg, Julie Farb, Janet Gillman, Shannon Gleeson, Kati Griffith, Quenton Herbert, Erin Johansson, Tamara Lee, Javier Ramirez, Christine Riordan, Palak Shah, and Marc Weinstein. Board members and RVP's present were Beverly Harrison and Jim Pruitt. Committee Chairs present were Virginia Doellgast, Joel Cutcher-Gershenfeld, Abigail Wozniak, Harry Katz, Taz Hurst, David Lewin, Bonnie Castrey, and Steve Sleigh. Officers present included Ryan Lamare, Bill Canak, Adrienne Eaton, Paul Clark, Wilma Liebman, and Andrew Weaver. Staff present included both Emily Smith and Bernadette Tiemann.

Approval of the Minutes—The motion to approve the minutes passed unanimously for both the board meeting in June of 2021 as well as the General Membership Board Meeting in June of 2021 with no revisions. Bill Canak moved the motion to approved and Beverly Harrison seconded the motion which passed unanimously.

Reports

Discussion of Meeting Registration for Speakers—Wilma Liebman, President, raised the question of charging speakers for meeting registration, specifically for practitioner speaker who intend only to attend the meeting to speak and participate at their session only and not attend other session on the meeting program. She reported that last year several meeting speakers questioned their need to pay for registration when their intention was to minimally participate in other ways. Wilma asked the board to consider alternative payment schemas, either to be considered on a case-by-case basis or some other system. After some discussion on the topic, it was tabled to hear from both the financial, membership, and program committees. After those reports were heard, the board continued discussion of this topic. Some ideas included: 1) a reduced one-day rate, perhaps $50, for virtual speakers, and/or 2) to publicize hardship scholarships for speakers. One board member thought that perhaps $50 was too low for a one-day rate for virtual conference registration and that we should focus on relaying the value of participating at the conference. Another board member offered that the goal of coming the meeting is building community and connection and dropping in to participate in only one session does not meet that goal. Many voiced support of hardship scholarships for those who could not otherwise financially participate but urged fiscal caution for enacting policies that the organization would not be able to financially sustain in the long run, considering that speakers far outweigh attendees. After divergent discussion, the topic was tabled for further examination and may again be considered at future board meetings.

Financial Report—LERA Secretary-Treasurer Andrew Weaver reports that had a slight decline in revenue in comparison to 2020. There is some softness around membership numbers which will discussed in more detail with the membership report and there are some strategies that have been proposed and built into the budget next year to address that. We had a small decline in revenue; however, expenses declined even more. We were able to realize an 11% net income up from a 6% net income in 2020, so there was in improvement there. Going forward, we expect our net income to again decline to 6% in 2022. There is currently high inflation, and we do need to make sure that our critical staff, Emily Smith and Bernadette Tiemann, who are doing an excellent job, are keeping pace. Sadly, we will not be increasing Emily's salary enough to keep pace with inflation, but the 2022 budget calls for a 3% raise for Emily Smith and a 6% raise for Bernadette Tiemann. We are also seeing higher expenses in publishing and mailing costs in 2022. With flagging membership, we are also increasing our membership promotion budget including direct mail and digital promotion to gain back necessary members to sustain the organization. All of this is going to create a tighter budget in 2022. Staff has done a good job holding the line and preventing pandemic-related losses as we struggle with not having in-person meetings and taking a more remote approach to the entire organization. We haven't seen the bottom fall out of membership and we've kept the organization together. Revenues from chapters have been really encouraging; there's been good chapter engagement. However, membership is dropping, and that is going to require investment and digital recruitment going forward. All of this is going to squeeze; that plus inflation over the next year will put our bottom line back into single digits in the budget for 2022. For the virtual meeting, we realized $104,000 in meeting revenue, which was right where we predicted we would. Contributions held steady from individuals at $35,000. The diversity and inclusion fund is comprised of two components: $10 from each full meeting registration on a four-year moving average and contributions to the Gershenfeld fund. Those two figures combined is about $9,067. Last year, after it was announced that the Gershenfeld fund was being repurposed to support diversity and inclusion our members responded generously with over $5,000 in contributions, up from $810 in the previous year. Emily Smith, LERA Executive Director, encourages the board to use the funds in ways that both meet the mission of increasing diversity and inclusion at LERA meetings and support the operating expenses of the organization, since the funds have been diverted from operating expense capital, for sustained financial stability. Tamara Lee motioned to accept the financial report and Bill Canak seconded the motion. The motion to accept the financial report from 2021 and the 2022 budget passed unanimously.

Membership Committee Report—David Lewin reports that Tazwell Hurst was recently appointed the second co-chair of this committee. The pandemic year one, 2020, was ironically good for membership, as membership increased. But 2021 membership is lower than 2020, but higher than 2019. In early 2022, a lag has begun, but we will work to increase membership throughout the year with a multi-faceted marketing promotion approach. The proportion of members that are represented by attorneys, government, management, and unions., has declined by about 10% since 2010. Neutral members have remained about the same. Academics, however, have increased their numbers. David Lewin makes this observation as a concern about its impact on practice point of view. He does not believe that LERA will ever be exclusively an academic organization but feels it's important to point this trend out. Wilma Liebman asked the question of what we should be doing to attract more members in the practitioner community. Adrienne Eaton had initiated some discussion last year, and there is a template letter included in the board packet that can be used by anyone on the board or membership committee as template language to reach out to members of management to invite them to join, and all members are invited to do so. Harry Katz mentions that it is important to continue connecting national members to our local chapters. Paul Clark mentions that Steve Greenhouse has been writing about the renewal of the "labor beat" among journalists, and that perhaps we could speak to Greenhouse about getting labor journalists more involved in LERA, especially in light of the fact that journalists can apply for press passes to our annual conferences. It was mentioned that Greenhouse will be involved at the next meeting, and there is another initiative in the works which may create a Labor Journalism Interest Section within LERA. At the LERA Presidential Address planned in 2022, Josh Eidelson, a Bloomberg reporter, will interview Wilma Liebman. Erin Johansson reports that unions may be interested in becoming more connected through the industry councils with LERA and that she will connect with Joel Cutcher-Gershenfeld on how to better implement this connection. Bill Canak thanks Harry Katz for bringing this up the important connection between national LERA and our chapters. Taz Hurst suggested we reach out to organizations like the AFL-CIO and ask them to promote LERA membership with LERA to their members. He mentioned the same strategy could be taken with other organizations he is familiar with including the CBTU APRI and the Machinists Union. Taz Hurst attends these conferences annually and may be able to promote LERA at those events. Tami Lee volunteers to ask members to find out their practitioner experiences of the meetings and of membership. Julie Farb comments that it might be effective to reach out to a group of international union research directors and invite them to connect with LERA. The AFL-CIO convention is coming up in June 12, 2022 in Philadelphia, and Julie Farb suggested this conference may be an excellent opportunity for LERA to have some sort of presence and promotion at the meeting. Wilma would like to follow-up with Julie Farb on these ideas. Taz Hurst mentions that the Machinists Union is having their conference in September. They have a training center in Southern Maryland, and they could reach out to their members though this venue. Emily Smith will follow-up with Taz following the Board meeting on how to connect Tax with LERA resources when he attends these conferences. Bonnie Castrey will work to connect local chapters with this effort. Erin Johansson asked about bringing back the National Policy Forum in DC. Wilma promises to work on these efforts with a goal to see increased membership by our June meeting 2022 particularly with practitioners. The membership committee report was adopted unanimously by the board. Peter Berg moved and Bill Canak seconded the motion.

LERA Annual Meeting Program Committee Report—Paul Clark reports that he is excited about the upcoming program and thanks the Vice Chairs Bill Spriggs, Bob Chiaravalli, and John Budd and the program committee members for their work. The theme of the program is "Elevating Voice and New Voices in the Workplace and Beyond." There were several sessions submitted on the meeting theme, and there are several strong plenaries planned, and one or two more that may still be announced. We are hoping that we get great attendance. Wilma Liebman thanks Paul Clark and his committee for such good work.

ILERA/LERA Program Committee Report—Harry Katz reports that he and Emily Smith are close to securing a contract with midtown Manhattan Hilton for the joint 2024 LERA/ILERA event that will take place the last week of June 2024. He is committed to creating a robust scholarship fund to support individuals to participate who could not otherwise from around the world including Southeast Asia, Africa, and Latin America. The initially proposed location in California did not pan out with the hotel, and the hotel in Manhattan not only has excellent facilities for our meeting but also seems to be offering us fairly preferential terms. We will report back at the next board meeting. We hope to attract 600-700 people at this exciting joint event.

LERA@ASSA Program Committee Report—Abigail Wozniak reports LERA@ASSA conducts 18 sessions in conjunction with the AEA/ASSA Annual Meetings, a tradition for LERA going back 74 years. In fact, LERA is one of the six founding members of the ASSA. These 18 sessions represent additional opportunities for LERA members to present their research and keep the issues of importance to LERA front and center in mainstream economics and with policy-makers. The 2021 program featured several important sessions with key speakers. To ensure that we have a sustainable model at this meeting, LERA requires that all those who wish to present a paper in a LERA session become a LERA member. Brad Hershbein was recently appointed as the second co-chair by Wilma Liebman. This year the committee has been re-furbished and is comprised of 16 strong and diverse committee members; it's an excellent committee. March 15th, 2022, is our session proposal deadline for our 2023 meeting in January; you are invited to submit your session proposals. Upcoming meetings are still all planned to be place-based meetings, though that could change. These include: 1) January 6-8, 2023 (Friday, Saturday & Sunday) New Orleans, LA, 2) January 5-7, 2024 (Friday, Saturday & Sunday) San Antonio, TX, and 3) January 3-5, 2025 (Friday, Saturday & Sunday) San Francisco, CA.

Development Committee Report—Harry Katz reports that development funds have contributed to the solid financial state of LERA reviewed by our secretary-treasurer. We raised $35,000 last year in individual contributions, and $40,000 from our major sustaining sponsors, Kaiser Permanente and Ford Motor Company. International Union, UAW is a new sustaining sponsor for LERA beginning in 2022 for $25,000. We are also building a quasi-endowment fund focusing on major gifts and final gifting, so please think about this in your estate planning. We are seeking additional sustaining sponsors, and if you have leads, Harry Katz and Jim Pruitt are happy to make the ask. Jim Pruitt made a personal request that each individual board member and officer consider donating a $1,000 each year, depending upon individual personal ability to do so. Jim also believes there could be interest from Kaiser Permanente managers in LERA membership and will connect with Emily Smith following the meeting to discuss further.

Nominating Committee Report—Steve Sleigh reported his slate of 31 potential candidates to the board for the June 2022 LERA election. His full committee, save Kris Rondeau who was unable to attend, met on Friday, February 4, 2022, and determined the slate that is contained in the board packet. The committee well-represented all the viewpoints of the organization. Wilma Liebman asked the board if they had any questions or had any comments after reviewing the slate and relayed that the board does have the authority to make changes to the slate before approving it. Liebman raised questions she had related to the slate of candidates: 1) two nominees in the neutral category are perhaps too similar to each other, and 2) a nominee in the management category is potentially being too busy accept. The slate passed unanimously without revision. Marc Weinstein motioned to approve the slate, and Bill Canak seconded the motion and it passed unanimously.

Interest Sections/Industry Councils Coordinating Committee Report—Joel Cutcher-Gershenfeld, reported the numbers of members who identify with the various LERA Interest Sections and Industry Councils. In this way, members are voting as to where their interests lie. Our challenge is how to appreciate that and how to expand it going forward. Momentarily, Paul Clark will present a new interest section charter for your review on pedagogy and teaching in our field, which we are all excited about. The second initiative is that there's organizing going on for a labor journalism/media industry council. It's still under discussion and yet to determine a name, but there are currently eight or nine alt media outlets that are unionized including some very well-known ones. And there are labor journalists still at major media outlets. It's likely that at the next annual meeting of the board, we may have a charter of that group coming. If you are interested in either of these initiatives, please contact Joel Cutcher-Gershenfeld, and he can point you to the individuals working on them. The third initiative is that we have identified the lowest-populated industry councils and see if we might be able to combine some of them and perhaps breathe new life into them. The real aim is for LERA to have communities of practice by interest and industry/sector, by which people can of course contribute to the program, contribute to webinars, and frankly be a voice for those industries or sectors in society. In the areas of construction, healthcare, and higher education, each have affiliates with other centers that deal with employment issues in those respective arenas, and that has been a powerful additional organizing feature.

Teaching Interest Section Report—Paul Clark reported that the idea of a LERA Teaching Interest Section arose at the June 2021 Annual Meeting. The Section organized a session on the program which reviewed innovation in the field and online teaching, and there was a good turnout. The group is continuing to move forward and it has two sessions on the upcoming June 2022 program. One of the things they did in preparation for the upcoming meeting was to send out a survey, asking people about the innovative things they have been doing in the classroom. Antone Aboud has been leading that, and Karla Katz has also been involved; they are both working through the responses to the survey. There has been plenty of interest, and those interested have shown initiative and follow-through, so I propose we take this forward. A motion to approve the Teaching Interest Section charter came from Joel Cutcher-Gershenfeld, was seconded by Adrienne Eaton, and was passed unanimously by the LERA Executive Board.

Diversity and Inclusion Report—Virginia Doellgast reported that we need to be making sure we are more effectively using the funding available to fund the people who need it to participate at LERA Annual Meetings. In trying to find a happy medium between making sure people have access, and not over-promising what our meager funding can afford, we have landed on placing information about diversity and inclusion funding for under-represented minorities on the session proposal submission forms for organizers to see when they submit their session proposals. These organizers are in the best seat to accurately assess and point us to speakers who may have a financial need and can alert the program committee and the diversity and inclusion committee to this. This information was not in the 2021 session proposal forms, but it is in the 2022 session proposal forms and will be there going forward. We still need to get this information reported to the diversity and inclusion committee for 2022 so they can review and approve any requests that have been made by session organizers. Additionally, there is another, stand-alone form that individuals can independently access, outside of the session proposal form, to request registration support. In 2021, 4 requests were made, and 4 requests were approved by this committee through that independent form. In 2022, no requests have yet been received. The committee has implemented a code of conduct that all meeting attendees are required to agree to, and meeting attendees are asked to report on their gender and race. Members are also asked to report on their gender and race. LERA has instituted language that specifies that those accepting awards must be free of harassment findings within a certain time period. Program Chairs are asked to consider diversity in their programs. In tracking sessions on the program that have a significant focus on equity and justice, based on race, gender, disability, and age, Doellgast reported 11 significant panels in both 2021 and 2022 which is an increase from prior years. President Liebman thanked the committee for their report and requested that they keep up the good work.

Editorial Committee Report—Ryan Lamare reports on the 2022 Research Volume "Industrial Relations and a Racial Reckoning" by Tamara Lee, Maite Tapia, Naomi Williams, and Sheri Davis Faulkner. This book is finishing up production with the majority of chapter in the production process. We anticipate being able to complete the volume and have everything ready to go and in the mail on time, and he thanked the editors for all their outstanding work. The 2023 Research Volume proposal is contained in the board packet for review "The Evolution of Workplace Dispute Resolution: An International Perspective," edited by Ariel Avgar, Cornell University; Deborah Hann, Cardiff University; Ryan Lamare, University of Illinois at Urbana-Champaign; and David Nash, Cardiff University. This proposed volume carries forward some trends that we would have seen through the LERA Research Volume process as far back as the late 90's where we had a volume on dispute resolution and workers' rights, edited by some folks here. We wanted to try to propose a volume that carries forward that conversation about the role of conflict in societies with a broader set of perspectives and twenty years later, considering cross-national and international components and also conversation about the role that conflict plays with major events and shocks that have occurred in the last twenty years, in particular the pandemic, the rise of #MeToo, and the role that conflict resolution programs play within the field of labor and employment relations. This proposal includes a variety of perspectives, international in particular. The editors have solicited chapter contributions that will go beyond the domestic focus that we tend to have in conflict research, and we have been able to secure chapters from authors focusing on the US, UK, Germany, Ireland, Australia, New Zealand, and others. Tamara Lee motioned to accept the proposal, and Beverly Harrison seconded the motion, and the motion passed unanimously.

National Chapter Advisory Council Report—William Canak, NCAC Chair, reports that after two years of pandemic-related challenges the LERA chapters are in exceptionally good shape. Many are up for chapter merit awards this year, and as you heard earlier, chapter dues payments have been great, which he believes is a mark of the health of local and state chapters. Thanks to Paul Clark, we have begun an initiative aimed at founding new chapters in several promising cities and states around the U.S. That initiative will be coming in the months ahead and we hope to give you progress updates in June this year. Many of you have seen evidence of chapter activity reported in the LERA eBulletins. This reporting confirms the energy and vitality of the local chapters. During the pandemic, the ways that they have innovated means they are engaging and serving their members. We have RVP election coming up for the East region. I'll note that all regional chapter vice presidents have served with distinction and really confirmed the excellence of this change to LERA's bylaws a number of years ago. In addition, LERA chapter liaisons have been reassigned this past year and are working, and are available to work, with chapter RVP's and individual chapters. The NCAC has had several virtual meetings in the last year which were engaging and productive. In June, we will conduct meetings of the NCAC, a chapter representatives meeting, and we'll again conduct our chapter administration workshop session. This year's session will feature Virginia LERA, a newer but highly productive chapter, and COLERA, which has been extremely active during the pandemic. Each chapter will talk about their approaches. Bill Canak ended his report by encouraging all LERA national members, and board members, to join their local chapter, as required by LERA bylaws. If anyone needs help connecting with a local chapter, Bill stands by to assist.

Next LERA Executive Board meeting will take place virtually on Wednesday, June 1, 2022, at 6-7:30 p.m. Eastern Time (New York Time).

Adjournment—The meeting was adjourned at 12:52 p.m. Eastern Time by President Wilma Liebman.

LERA Executive Board Meeting Minutes Wednesday, June 1, 2022, 5:30–7 p.m. Eastern Time Via Videoconference

Call to Order—The meeting was called to order at 5:30 p.m. by Wilma Liebman, President. Present at the meeting were board members Daniel Altchek, Peter Berg, Julie Farb, Janet Gillman, Shannon Gleeson, Kati Griffith, Quenton Herbert, Erin Johansson, Tamara Lee, Javier Ramirez, Christine Riordan, Palak Shah, and Marc Weinstein. Board members and RVP's present were Beverly Harrison and Jim Pruitt. Committee Chairs present were Virginia Doellgast, Joel Cutcher-Gershenfeld, Abigail Wozniak, Harry Katz, Taz Hurst, David Lewin, Bonnie Castrey, and Steve Sleigh. Officers present included Ryan Lamare, Bill Canak, Adrienne Eaton, Paul Clark, Wilma Liebman, and Andrew Weaver. Staff present included both Emily Smith and Bernadette Tiemann.

Approval of the Minutes—The motion to approve the minutes passed unanimously for board meeting in February of 2022 with no revisions.

LERA 76th Annual Meeting Update—Harry Katz reports that the meeting that will occur in 2024. Harry began visiting this board a few years ago to coordinate the ILERA triennial world congress synchronously with the LERA 76th Annual Meeting. The latest development is that Javier Ramirez from the FMCS proposes that in 2024 we also coordinate this meeting jointly with the FMCS National Labor-Management Conference. This meeting is held every other year and draws a substantial crowd of labor and management practitioners. The board discussed, offered their views, and expressed enthusiasm for this joint venture. The venue for the LERA 76th Annual Meeting, the Hilton Midtown Manhattan, is of a scale that will accommodate all three organizations. The plans are underway, the date is June 26-30, 2024, and the event should be engaging with something to appeal to all constituencies. There will be particular events that will be still occur, like General Membership Meeting for LERA and ILERA, Presidential Luncheon for LERA, there is plenty of session space and common space for attendees to interact and network. Javier Ramirez at the FMCS reports that LERA needs to both maintain academic elements of the conference and to generate revenue. Neither of these issues are concerns for the FMCS. The FMCS does have their conference every two years and would like to have this collaboration recur every other year, with a memorandum of understanding between our two organizations. This gives us a significant opportunity to increase LERA membership, drive revenue for LERA, etc. There are many opportunities for efficiency and redundancy reduction, including: key speakers (such as the Secretary of Labor Marty Walsh), contributors who would need to contribute to a single event, shared administration and meeting platforms, etc. We have the ability to work closely on the content that overlaps and keep other content separate. The main question is: Why have we not done this sooner? Here is a chance to do it, and even if we experience a learning curve, and we will continue to improve the process for 2026 and beyond, and this this the main event to attend in the field of LER. At this point we are committing to having a joint meeting in 2024. If all goes well, we can continue doing this in 2026 and going forward. Harry Katz mentions that LERA has many academics, and FMCS has many practitioners, and this gives us an opportunity to enhance the experience of the parties involved on both sides, somewhat broader to where LERA operated 20-30 years ago. If we have more practitioners at the meeting, it can only elevate the event. To tighten the bond even further, Joel Cutcher-Gershenfeld suggests that we might have FMCS mediators tied to various LERA Industry Councils and have sessions at the meeting featuring the mediators. The board voted and the vote was unanimous to move ahead with the joint collaboration. Additionally, a memorandum of understanding will be developed between our two organizations (LERA and the FMCS) to formalize this relationship.

Reports

Financial Report—LERA Secretary-Treasurer Andrew Weaver reports that the final 2021 audit has not been completed but we should be getting the final 2021 numbers soon. This is hopefully the last year of our virtual conference. Revenues declined a small amount, but expenses declined by more. The bottom line is that we were in the black in 2021 and we cleared a percentage of net income that is consistent of our past experience under Emily Smith's directorship. On the whole, 2021 ended up in a good place. Going forward, the revised 2022 budget, we have had a very good story on sponsorship thank you to UAW, Ford, and Kaiser Permanente and the Alliance of Healthcare Unions. We do need to make some investments. First, salaries increased by one percent from 2021 to 2022. In coming years, we may look at increasing salaries. There is extra investment in the 2022 budget for extra-help in the fall during the busy season. We also have investments planned for our library digital subscriptions, and chapter websites, and other matters. In 2022, we are expecting to clear about 6% net income or a little over $20,000 and stay in the black. On the whole, LERA is in good shape, and when we ratchet back up to in-person conferences we will be in a good footing. The LERA board accepted the financial report.

Membership Committee Report—David Lewin reports that membership numbers were flagging in 2021, but that those numbers have been shored up in 2022 due to the new major sustaining sponsors, which carries a component of membership as well. You will notice that the price of membership has stayed the same for about a decade, which is remarkable, especially when benchmarked to other organizations. We have held the line on dues. This, and our positive financial report, is all do to a constrained financial approach. Wilma Liebman says that her regret over the last year has been being unsuccessful with recruiting more practitioners in LERA, and she would love to see this portion of the membership grow. Perhaps the collaborative event planned for 2024 will help this. Kevin Legel has begun publicizing an opportunity for their staff to have discounted memberships at LERA through Ford's major sustaining sponsorship.

LERA Annual Meeting Program Committee Report—Bill Spriggs reports that due to technical glitches, today's earlier program committee meeting will be repeated in late June or early July. The theme for the LERA 75th Annual Meeting, June 1-4, 2023, in Detroit, Michigan is planned to be: "Workers and American Democracy: Challenges Ahead, Lessons from Before." Bill shares that he has been personally troubled by the demise of democracy globally, and he encourages papers or sessions where we can appreciate this, and think about contributions of labor to the legal democracy in the United States, its impact on American workers and global workers and on American foreign policy, etc. We hope to examine the attacks on democracy and related challenges, but also to look to the future and point to people who are engaging in collective action and education around organizing. We are currently in a moment where new types of collective action are taking place, and it is appropriate to pay attention to historical contributions of the labor movement to the quality of work, life, and employee involvement, and hypotheses that if people had more of a say at work, it would roll over into civic life and improve democracy, an idea which has been supported by a number of studies.

Development Committee Report—Harry Katz reports that LERA is in good shape on the development front as we now have three distinct major sustaining sponsors including UAW, Ford Motor Co., and Kaiser Permanente and the KP Alliance of Healthcare Unions. Please know that when you and others make contributions to LERA, you can always target gifts to specific worthy causes, such as various funds. We are also building a quasi-endowment fund focusing on major gifts and final gifting, so please think about this in your estate planning or major gift. We are seeking additional sustaining sponsors, and if you have leads, Harry Katz and Jim Pruitt are happy to make the ask. Jim Pruitt made a personal request that each individual board member and officer consider donating a $1,000 each year, depending upon individual personal ability to do so.

Interest Sections/Industry Councils Coordinating Committee Report—Joel Cutcher-Gershenfeld reported that the industry councils and interest sections will meet Friday at 3 pm and they will take a deeper dive into the information that is contained in the report attached in the board packet. Joel will be asking the section and council leaders about potential rebranding. Secondly, we have had low involvement in the airline and utility councils and those may need to be retired. Joel looked through the program with an eye towards content about what sessions speak to issues of interest to various sections and councils, and there were several sessions of interest to many of them. At our meeting on Friday, we will be discussing the possibility of exploring new sections or councils on the areas that were most noted on the program. The current leadership on councils and sections are listed in the report and we would love to correspond a federal mediator to each of these groups. The media award committee has been working to establish a labor journalism group and there will be a panel discussion by former award winners on the program this year. We feel we are at a point at reaching out to these individuals and establishing a formal network of labor journalists. Quite a few of these labor journalists are organizing themselves into unions. Steven Greenhouse has a list of a few dozen labor journalists that he is in touch with and he is willing to share this list with the group. Javier Ramirez looks forward to having that conversation about mediators and also has some suggestions about the airline industry council. When people speak to labor journalists, they should be given the heads up about LERA and tell them about the new journalism group. The newest interest section is on teaching and pedagogy, which Paul Clark has played a large role in creating.

Diversity and Inclusion Report—Christine Riordan reports that we have added in gender and ethnicity demographics into multiple touchpoints of LERA processes, and we have collected about half of our members' demographics and about two-thirds of our meeting registrant's demographics. One way in which we are currently using this is to work with the program committee. Organizers have the ability to give information to the program committee on the diversity and inclusion of their session. We also have a fund to build diversity and inclusion in various ways. We have begun to lay a foundation for collecting information and building these efforts into our processes. We are looking at giving membership to those who are approved for meeting registration. We are also looking at providing childcare resource information to LERA meeting attendees. The committee discussed LERA's meeting locations where issues of women's healthcare is de-prioritized. The committee is supportive of looking into this further and seeing what other organizations are doing and seeing whether any action by LERA is feasible. Harry Katz offered that the committee is not yet coming forward with a proposal to alter how LERA does our site selections, but please do keep in mind that LERA does strive to do two things: we want to be in a union hotel and we want to be in a location that has a local LERA chapter, and that he would caution against reducing our support or services to those chapters. Bill Canak expresses that he supports meeting in locations where LERA local chapters exist or are forming and that this should be an important consideration when reviewing where to conduct LERA meetings.

Editorial Committee Report—Ryan Lamare reports that the LERA/ILR Review process had ten submissions after the last June conference. Of those, three moved into the review process, and two remain in the review process. Later this summer, we will discuss how to make this process more efficient and get the papers through the process in a more compressed timeline. The LERA 2022 Research Volume "Industrial Relations and a Racial Reckoning" by Tamara Lee, Maite Tapia, Naomi Williams, and Sheri Davis Faulkner is almost out the door; we are proud of the volume and will be delighted to see it in print. Tamara Lee and Matie Tapia who are on today's call have done a tremendous amount on this book. This book will be on the shelves in the next few months. The 2023 Research Volume "The Evolution of Workplace Dispute Resolution: An International Perspective," edited by Ariel Avgar, Cornell University; Deborah Hann, Cardiff University; Ryan Lamare, University of Illinois Urbana-Champaign; and David Nash, Cardiff University was approved in February. For both of these volumes, there are sessions on the program to take a look at this content. The conflict resolution volume is moving along well. At our next meeting, we will likely ask for your approval of the 2024 research volume which will be an update of the 2013 research volume on collective bargaining under duress. This will be edited by Howard Stanger, Paul Clark, and John Delaney.

National Chapter Advisory Council Report—William Canak, NCAC Chair, reports that LERA chapters are in good shape. Chapters are active and coming out of the pandemic era, chapter dues payments have been great. We have new chapters that have formed (Virginia LERA), and thanks to Paul Clark, we have begun an initiative aimed at founding new chapters in several promising cities and states around the U.S. such as southern Florida and in Mississippi and Louisiana. That initiative will be coming in the months ahead and we hope to give you progress updates in June this year. Many of you have seen evidence of chapter activity reported in the LERA eBulletins. This reporting confirms the energy and vitality of the local chapters. We have RVP election coming up for the East region. I'll note that all regional chapter vice presidents have served with distinction and really confirmed the excellence of this change to LERA's bylaws a number of years ago. In addition, LERA chapter liaisons have been reassigned this past year and are working, and are available to work, with chapter RVP's and individual chapters. The NCAC has had several virtual meetings in the last year which were engaging and productive. In June, we will conduct meetings of the NCAC, a chapter representatives meeting, and we'll again conduct our chapter administration workshop session. This year's session will feature Virginia LERA, a newer but highly productive chapter, and COLERA, which has been extremely active during the pandemic. Each chapter will talk about their approaches. Bill Canak ended his report by encouraging all LERA national members, and board members, to join their local chapter, as required by LERA bylaws. If anyone needs help connecting with a local chapter, Bill Canak, Bonnie Castrey, and the NCAC members stand by to assist.

Some of the attachments include the LERA@ASSA 2023 program, the 2022 LERA award winners, the LERA 2021 election winners. Paul Clark reports that the program committee feels great about it's theme of Elevating Voice and New Voices in the Workplace for the current annual meeting. It was mentioned today that one of the beautiful things about the virtual meeting is that people can continue to register right up to and through the meeting. The opening session tomorrow is with the U.S. Secretary of Labor Marty Walsh. Wilma Liebman did invite him to come back next year in person. Wilma Liebman thanks the staff for their dedication. Deborah Mueller came to a conference a few years ago and left with a mission to regenerate the Western NY LERA (Buffalo) chapter, and then the pandemic changed everything. As things have gotten back to normal, she had other circumstances that have precluded this activity again. However, she can see the finish line and is hoping that we can get back to this goal of revitalizing this chapter and will send out some thoughts and suggestions to potential volunteers and reach out to Virginia LERA and others for suggestions and best practices on chapter-building. Further, Deborah reports seeing large labor strikes resulting in big changes for both workers and communities, and that this is a good time for the Buffalo Chapter to revive. Bill Canak encourages her to come to the chapter workshop featuring this exact topic and to additionally contact the NCAC for additional support.

Next LERA Executive Board meeting will take place virtually on February 21, 2023, at Noon Eastern Time, via video conference.

Adjournment—The meeting was adjourned at 5:51 p.m. Eastern Time by President Wilma Liebman.

LERA General Membership Meeting Minutes Friday, June 3, 2022, 3:45 p.m. Eastern Time Via Videoconference

Call to order—Wilma Liebman opened the meeting at 3:45 pm Eastern Time, and introduced Paul Clark, LERA President Elect and at the end of the LERA 74th Annual Meeting, officially the LERA President.

Financial Report—Andrew Weaver, Secretary-Treasurer, reported that the LERA is in good shape financially since expenses were held down in relation to decreased income in 2021 and 2022. However, we are looking forward to returning to the full financials from pre-pandemic era with the return of in-person meetings in 2023.

Membership Committee Report—David Lewin, Co-Chair, reported membership is in good shape. Academics comprise about 55% of our membership and other groups account for the rest. Given the registration for this particular conference, which is around 560 people, Lewin felt we could look forward to additional membership increases. This 2021 increase is due to a new major sustaining sponsor, United Auto Workers, and our other major sustaining sponsors (Kaiser Permanente, Ford, etc.) If we were to count local chapter members in our numbers, we would be closer to around 4,000 members, of course.

Special Announcement—Javier Ramirez and Harry Katz. Harry Katz is currently the president of ILERA in the midst of a 3-year term. At the end of his term, he will organize a world congress of ILERA which will happen in the US. Two years ago, it was decided that ILERA and LERA would conduct our meetings synchronously, and most recently Javier Ramirez suggested that we also synchronously conduct the FMCS national labor-management conference. The LERA Executive Board agreed to that as did the ILERA executive board, and so we are now planning a meeting the last week of June 2024 which will lead to an enriched program, and it should help expand LERA membership among the practitioner community.

Javier Ramirez had approached LERA and LERA responded that they needed to keep our sessions for our academics. With the FMCS, they run into issues with money every year, and it also benefits with LERA, and the thought is that every two years we could continue to meet together. We will make sure there are protections for LERA in place with a memorandum of understanding in case future administrations decide to move in a different direction. In the future, the FMCS will be collaborating with us, and there are overlays in capacities with similar content. And, it's a way to build the membership for LERA, there are pricing structures that would give discounts to LERA members, and it makes sense for the labor-management community at large. Networking is going to increase, and partnership between academics and practitioners will increase. The meeting will take place at the Midtown Manhattan Hilton on 6th Street, and we will be able to socially interact and network as community.

Development Committee Report—Harry Katz and Jim Pruitt, Co-Chairs, reported contributions are healthy as are our finances more generally. As was already mentioned by David, we benefit from having three distinct contributions (one from UAW, one from Ford, and one from KP and the KP Alliance of Health care Unions). We also have solid annual fund-raising drive. Keep in mind you can give money in either a restricted manner, or something that is dear to your heart like helping graduate students attend these meetings, or for worker centers attend our meetings. Jim Pruitt reports that this general membership meeting is very positive and makes a long-term LERA members feel that we are moving in a positive direction. Jim asked anyone in our general membership to contact the LERA staff and make a personal donation.

Rose Batt weighed in that the diversity within this organization has really grown and that she had the good fortune of not knowing most of the people at this conference, and additionally feels like there is a real momentum at the organization. There are all sorts of creative things that we can do to keep this thing rolling. Rose was impressed by what she has seen and appreciates the changes and differences. Wilma Liebman mentions that some of it is owed to the unique dynamism of the moment that we are in currently.

Editorial Committee Report—Ryan Lamare reports that the LERA/ILR Review special section had ten papers initially submitted and that there were two still in the running. For this year, please submit your paper by June 30, and Ryan promises to try to get an expedited review as possible. A couple of other updates, LERA's 2022 Research Volume "Industrial Relations and a Racial Reckoning" will be released shortly and the editorial committee is proud of this book. Our 2023 Research Volume will be on conflict resolution and will highlight the current system of conflict management in various countries and across various industries. That will be edited by Ariel Avgar, Debra Hahn, Ryan Lamare, and David Nash, and there will be a session on the program tomorrow highlighting these topics.

Program Committee Report—Bill Spriggs, LERA President Elect and Program Committee Chair reports that for the LERA 75th Annual Meeting, the theme will be workplace democracy, a topical and timely topic, especially with the concerns worldwide about the erosion of democratic institutions. The meeting will take place in-person in Detroit, Michigan in early June. Please submit your session proposals to the program committee by November 15, 2022.

Diversity and Inclusion Committee Report—Christine Riordan reports that the Diversity and Inclusion committee has been working to promote diversity and inclusion within our membership and at our meetings. Our sexual harassment policy is part of that, and last year we instituted a number of ways to collect demographic information on our meeting attendees and with our members. We have been working with the program committee to increase diversity of speakers. The Gershenfeld fund has been repurposed to support diversity at LERA annual meetings. There are a number of ideas that the committee is considering for future work.

National Chapter Advisory Council Report— Bill Canak, Chair, and Bonnie Castrey, Vice Chair of the National Chapter Advisory Council report that the state of the chapters is excellent. We have a number of chapters receiving chapter merit awards and outstanding chapter recognition. We are working with the FMCS to put together an organizing committee in Miami-South Florida area. We are also working with the FMCS to establish a new chapter in the southern Mississippi/New Orleans area. Those of you who read the bi-monthly LERA e-newsletter, please note that LERA is covering LERA chapter events, which helps us become a more cohesive organization. The vast majority of local chapter members are not academic, and the various activity that goes on in the local chapter, and you will see that activity is pragmatic and practitioner-oriented. Many of the changes made in the organization help support this increased weight of practitioners in the organization. We have two candidates running in the election this year for the East RVP. We have three chapter events happening at this conference this year, and much of the credit for these being successful and well-attended goes to the LERA staff and to the chapters. The chapter administration workshop will feature COLERA and Virginia LERA, our newest chapter. Since the meeting in 2016 in Minnesota, the bylaws of LERA have required all national members to be a member of a local chapter, and if you look into your own status and you see that you are not a local chapter member, please contact Bill Canak or LERA staff and they will help you find a local chapter. Wilma Liebman found it interesting and encouraging that LERA has chapters arising in the south. She further commented that LERA is a unique forum where all perspectives intersect.

Industry Councils and Interest Sections Coordinating Committee Report—Joel Cutcher-Gershenfeld, Chair, reported that we have eight established interest sections and one newly formed section, the Teaching Interest Section. We also have eight established industry councils and three new ones under discussion. Many of these groups are represented on the annual meeting program. When LERA members join or renew their membership they indicate which interest sections they want to affiliate with, and which industry councils, if any, with which they want to affiliate and he reported on current numbers in all these LERA groups.

LERA Awards Committee Report—Janice Bellace and Rose Batt, Co-Chairs, reported on the LERA 2022 Awards Winners. The capstone Lifetime Achievement Award was awarded to Bernard Anderson and Lisa Lynch. The Academic Fellows were Sarosh Kuruvilla, Chris Tilly, Françoise Carré, and Douglas Kruse. Practitioner Fellows were Meeta Bass, Erin Johansson, Mickey Brock, and Sheila Mayberry. The Susan C. Eaton Outstanding Scholar-Practitioner Award was conferred on Steven Greenhouse. The Myron C. Taylor Management Award was not conferred in 2022. The John T. Dunlop Outstanding Scholar Awards were given to Rachel Aleks and Chiara Benassi. The LERA Outstanding Practitioner Award was awarded to Owen Herrnstadt. The James G. Scoville Best International/Comparative IR/HR Paper Award was received by Chunyun Li. The Kenneth May Media Award was bestowed on Eli Rosenberg, Margot Roosevelt, Katie Johnston, and Michael Sainato. This year's Thomas A. Kochan and Stephen R. Sleigh Best Dissertation Award went to Debbie Goldman. The Susan C. Eaton Research Grant award was given to Jasmine Platt, and Ron Schatz. The DCDC Best Student Paper award winner was Jianxuan Lei.

Future LERA Meetings—To formally announce the dates of future meetings, we will be in Detroit in 2023 June 1-4, 2023 at the MotorCity Hotel and Casino. The LERA@ASSA 2023 will be in New Orleans, January 6-8, 2023, and the 2024 annual meeting will take place in New York City in Midtown Manhattan at the Hilton.

Presidential Handoff—Wilma Liebman formally handed over the gavel to Paul Clark, who will be assuming the mantle of LERA President at the end of this meeting. Wilma congratulated Paul on putting together this year's program. Paul Clark reported that it has been a pleasure to work with Wilma over the last year, and that she would be a tough act to follow. Paul shared that LERA has been a very important part of his professional life, and that he would his best to make sure that LERA is in good shape at the end of his presidential year.

Last Announcements—Tomorrow morning at 10 a.m. Eastern, LERA will kick off with the opening plenary on supply chains moderated by Tom Kochan followed by a full day of breakouts and sessions, and the presidential event which will feature our two lifetime achievement award winners.

Audit Report for 2020 and 2021

Labor and Employment Relations Association

Champaign, Illinois

Financial Statements

For the Years Ended December 31, 2021 and 2020

Feller & Kuester CPAs LLP
Certified Public Accountants
806 Parkland Court
Champaign, IL 61821
217-351-3192

TABLE OF CONTENTS

Feller & Kuester CPAs LLP

Tax - Audit - Bookkeeping
806 Parkland Court - Champaign, Illinois 61821
Phone - (217) 351-3192 Fax - (217) 351-4135 Email - neal@fellerkuester.com

INDEPENDENT AUDITOR'S REPORT

To the Board of Directors of
Labor and Employment Relations Association
Champaign, Illinois

Opinion

We have audited the accompanying financial statements of the Labor and Employment Relations Association (a nonprofit organization), which comprise the statements of financial position as of December 31, 2021 and 2020, and the related statements of activities, functional expenses, and cash flows for the years then ended, and the related notes to the financial statements.

In our opinion, the financial statements referred to above present fairly, in all material respects, the financial position of the Labor and Employment Relations Association as of December 31, 2021 and 2020, and the changes in its net assets and its cash flows for the years then ended in accordance with the accounting principles generally accepted in the United States of America.

Basis for Opinion

We conducted our audits in accordance with auditing standards generally accepted in the United States of America. Our responsibilities under those standards are further described in the Auditor's Responsibilities for the Audit of the Financial Statements section of our report. We are required to be independent of the Labor and Employment Relations Association and to meet our other ethical responsibilities in accordance with the relevant ethical requirements relating to our audits. We believe that the audit evidence we have obtained is sufficient and appropriate to provide a basis for our audit opinion.

Responsibilities of Management for the Financial Statements

Management is responsible for the preparation and fair presentation of the financial statements in accordance with accounting principles generally accepted in the United States of America, and for the design, implementation, and maintenance of internal control relevant to the preparation and fair presentation of financial statements that are free from material misstatement, whether due to fraud or error.

In preparing the financial statements, management is required to evaluate whether there are conditions or events, considered in the aggregate, that raise substantial doubt about the Labor and Employment Relations Association's ability to continue as a going concern within one year after the date that the financial statements are available to be issued.

Auditor's Responsibilities for the Audit of the Financial Statements

Our objectives are to obtain reasonable assurance about whether the financial statements as a whole are free from material misstatement, whether due to fraud or error, and to issue an auditor's report that includes our opinion. Reasonable assurance is a high level of assurance but is not absolute assurance and therefore is not a guarantee that an audit conducted in accordance with generally accepted auditing standards will always detect a material misstatement when it exists. The risk of not detecting a material misstatement resulting from fraud is higher than for one resulting from error, as fraud may involve collusion, forgery, intentional omissions, misrepresentations, or the override of internal control. Misstatements are considered material if there is a substantial likelihood that, individually or in the aggregate, they would influence the judgment made by a reasonable user based on the financial statements.

In performing an audit in accordance with generally accepted auditing standards, we:

- Exercise professional judgment and maintain professional skepticism throughout the audit.
- Identify and assess the risks of material misstatement of the financial statements, whether due to fraud or error, and design and perform audit procedures responsive to those risks. Such procedures include examining, on a test basis, evidence regarding the amounts and disclosures in the financial statements.
- Obtain an understanding of internal control relevant to the audit in order to design audit procedures that are appropriate in the circumstances, but not for the purpose of expressing an opinion on the effectiveness of the Labor and Employment Relations Association's internal control. Accordingly, no such opinion is expressed.
- Evaluate the appropriateness of accounting policies used and the reasonableness of significant accounting estimates made by management, as well as evaluate the overall presentation of the financial statements.
- Conclude whether, in our judgment, there are conditions or events, considered in the aggregate, that raise substantial doubt about the Labor and Employment Relations Association's ability to continue as a going concern for a reasonable period of time.

We are required to communicate with those charged with governance regarding, among other matters, the planned scope and timing of the audit, significant audit findings, and certain internal control related matters that we identified during the audit.

Feller & Kuester CPAs LLP

Feller & Kuester CPAs LLP
Champaign, Illinois

September 22, 2022

LABOR AND EMPLOYMENT RELATIONS ASSOCIATION
STATEMENTS OF FINANCIAL POSITION
DECEMBER 31, 2021 AND 2020

	2021	2020
ASSETS		
Current Assets		
Cash and Cash Equivalents	$ 394,971	$ 325,553
Investments	85,955	91,321
Accounts Receivable, Net	1,406	-
Prepaid Expenses	5,626	8,617
Inventory	6,372	6,685
Total Current Assets	494,330	432,176
Property and Equipment, Net	3,292	2,728
TOTAL ASSETS	$ 497,622	$ 434,904
LIABILITIES AND NET ASSETS		
Current Liabilities		
Accounts Payable	$ -	$ 750
Accrued Liabilities	39,661	42,055
Dues Collected in Advance	75,917	72,986
Subscriptions Collected in Advance	7,837	11,418
Deferred Chapter Dues	5,200	7,500
Other Deferred Revenue	54,550	23,175
Total Current Liabilities	183,165	157,884
Net Assets		
Without Donor Restrictions		
Designated by the Board		
Susan C. Eaton Scholar-Practitioner Memorial Fund	20,779	20,779
Kochan-Sleigh Best Dissertation Award Fund	36,439	37,439
Jim Scoville Award Fund	2,796	3,296
John T. Dunlop Public Policy Fund	18,326	19,005
Gladys and Water Gershenfeld Diversity & Inclusion Fund	16,585	17,147
Quasi-Endowment Fund	45,462	49,190
Total Designated by the Board	140,387	146,856
Undesignated	145,194	107,133
Total Without Donor Restrictions	285,581	253,989
With Donor Restrictions	28,876	23,031
Total Net Assets	314,457	277,020
TOTAL LIABILITIES AND NET ASSETS	$ 497,622	$ 434,904

See Accompanying Notes.

LABOR AND EMPLOYMENT RELATIONS ASSOCIATION
STATEMENT OF ACTIVITIES
FOR THE YEAR ENDED DECEMBER 31, 2021

	Without Donor Restrictions			With Donor	
	Undesignated	Designated	Total	Restrictions	Total
Support and Revenues					
Membership Dues	$ 148,934	$ -	$ 148,934	$ -	$ 148,934
Meeting Income	104,141	-	104,141	5,645	109,786
Organization Dues & Sponsorships	56,029	-	56,029	550	56,579
Subscriptions	15,348	-	15,348	-	15,348
Chapter Fees	12,175	-	12,175	-	12,175
Member Contributions	35,174	-	35,174	-	35,174
Royalties	6,463	-	6,463	-	6,463
Ad Income	10,230	-	10,230	-	10,230
Publications	1,272	-	1,272	-	1,272
Administrative Fees	1,535	-	1,535	-	1,535
Investment Income	(932)	(4,434)	(5,366)	-	(5,366)
Interest Income	310	-	310	-	310
Net Assets Released from Designation	2,385	(2,035)	350	(350)	-
Total Support and Revenues	393,064	(6,469)	386,595	5,845	392,440
Expenses					
Program Services					
General	217,170	-	217,170	-	217,170
Meetings	10,120	-	10,120	-	10,120
Publications	36,522	-	36,522	-	36,522
Supporting Services					
Management and General	77,543	-	77,543	-	77,543
Membership Development	13,648	-	13,648	-	13,648
Total Expenses	355,003	-	355,003	-	355,003
Change in Net Assets	38,061	(6,469)	31,592	5,845	37,437
Net Assets, Beginning of Year	107,133	146,856	253,989	23,031	277,020
Net Assets, End of Year	$ 145,194	$ 140,387	$ 285,581	$ 28,876	$ 314,457

See Accompanying Notes.

LABOR AND EMPLOYMENT RELATIONS ASSOCIATION
STATEMENT OF ACTIVITIES
FOR THE YEAR ENDED DECEMBER 31, 2020

	Without Donor Restrictions			With Donor	
	Undesignated	Designated	Total	Restrictions	Total
Support and Revenues					
Membership Dues	$ 149,333	$ -	$ 149,333	$ -	$ 149,333
Meeting Income	109,655	-	109,655	9,060	118,715
Organization Dues & Sponsorships	62,196	-	62,196	460	62,656
Subscriptions	16,031	-	16,031	-	16,031
Chapter Fees	10,328	-	10,328	-	10,328
Member Contributions	44,672	33,584	78,256	-	78,256
Royalties	6,549	-	6,549	-	6,549
Ad Income	9,880	-	9,880	-	9,880
Publications	100	-	100	-	100
Administrative Fees	520	-	520	-	520
Investment Income	(4,211)	(14,181)	(18,392)	-	(18,392)
Interest Income	460	-	460	-	460
Net Assets Released from Designation	5,514	(5,514)	-	-	-
Total Support and Revenues	411,027	13,889	424,916	9,520	434,436
Expenses					
Program Services					
General	219,005	-	219,005	-	219,005
Meetings	31,796	-	31,796	-	31,796
Publications	41,476	-	41,476	-	41,476
Supporting Services					
Management and General	73,798	-	73,798	-	73,798
Membership Development	11,483	-	11,483	-	11,483
Total Expenses	377,558	-	377,558	-	377,558
Change in Net Assets	33,469	13,889	47,358	9,520	56,878
Net Assets, Beginning of Year	73,664	132,967	206,631	13,511	220,142
Net Assets, End of Year	$ 107,133	$ 146,856	$ 253,989	$ 23,031	$ 277,020

See Accompanying Notes.

LABOR AND EMPLOYMENT RELATIONS ASSOCIATION
STATEMENT OF FUNCTIONAL EXPENSES
FOR THE YEAR ENDED DECEMBER 31, 2021

		MEETINGS			PUBLICATIONS			SUPPORTING SERVICES		
	General	Annual Conference	ASSA Meeting	Other Meetings	Perspectives	Research Volume	Other Publications	Management & General	Membership Development	Total
Compensation	$ 137,638	$ -	$ -	$ -	$ -	$ -	$ -	$ 30,943	$ -	$ 168,581
Payroll Taxes & Fringes	60,442	-	-	-	-	-	-	11,464	-	71,906
Contract Labor	19,090	-	-	-	-	-	-	-	-	19,090
Depreciation	-	-	-	-	-	-	-	2,079	-	2,079
Insurance	-	-	-	-	-	-	-	1,339	-	1,339
Bank and Service Charges/Fulfillment	-	-	-	-	-	-	-	15,698	-	15,698
Promotion	-	-	-	-	-	-	-	-	4,595	4,595
Postage/Freight	-	-	-	-	1,896	2,301	-	415	-	4,612
Accounting/Auditing	-	-	-	-	-	-	-	6,777	-	6,777
Printing and Production	-	-	-	-	8,498	7,674	-	-	-	16,172
Services	-	10,000	-	-	7,423	8,108	595	-	-	26,126
Other Publication Costs	-	-	-	-	27	-	-	-	-	27
Meals/Receptions	-	-	-	-	-	-	-	-	-	-
Travel	-	-	-	-	-	-	-	-	-	-
Other Meeting Expenses	-	-	120	-	-	-	-	-	-	120
Computer Supplies/Services	-	-	-	-	-	-	-	3,859	-	3,859
Website	-	-	-	-	-	-	-	1,845	-	1,845
Office Supplies	-	-	-	-	-	-	-	1,858	-	1,858
Student & Member Awards	-	-	-	-	-	-	-	-	4,784	4,784
Fundraising Expense	-	-	-	-	-	-	-	-	943	943
Duplicating Expense	-	-	-	-	-	-	-	1,156	-	1,156
Chapter Expenses	-	-	-	-	-	-	-	-	3,316	3,316
Other Committee Expenses	-	-	-	-	-	-	-	-	10	10
Miscellaneous Office	-	-	-	-	-	-	-	110	-	110
Total	$ 217,170	$ 10,000	$ 120	$ -	$ 17,844	$ 18,083	$ 595	$ 77,543	$ 13,648	$ 355,003

See Accompanying Notes.

LABOR AND EMPLOYMENT RELATIONS ASSOCIATION
STATEMENT OF FUNCTIONAL EXPENSES
FOR THE YEAR ENDED DECEMBER 31, 2020

		MEETINGS			PUBLICATIONS			SUPPORTING SERVICES		
	General	Annual Conference	ASSA Meeting	Other Meetings	Perspectives	Research Volume	Other Publications	Management & General	Membership Development	Total
Compensation	$ 143,096	$ -	$ -	$ -	$ -	$ -	$ -	$ 28,191	$ -	$ 171,287
Payroll Taxes & Fringes	57,459	-	-	-	-	-	-	10,681	-	68,140
Contract Labor	18,450	-	-	-	-	-	-	-	-	18,450
Depreciation	-	-	-	-	-	-	-	1,639	-	1,639
Insurance	-	-	-	-	-	-	-	892	-	892
Bank and Service Charges/Fulfillment	-	-	-	-	-	-	-	16,127	-	16,127
Promotion	-	-	-	-	-	-	-	-	-	-
Postage/Freight	-	-	90	-	2,124	2,600	-	387	-	5,201
Accounting/Auditing	-	-	-	-	-	-	-	6,707	-	6,707
Printing and Production	-	-	171	-	8,460	7,305	18	-	-	15,954
Services	-	14,000	-	-	11,738	8,605	552	-	-	34,895
Other Publication Costs	-	-	-	-	74	-	-	-	-	74
Meals/Receptions	-	2,000	10,110	-	-	-	-	-	-	12,110
Travel	-	1,267	2,239	539	-	-	-	-	-	4,045
Other Meeting Expenses	-	1,065	315	-	-	-	-	-	-	1,380
Computer Supplies/Services	-	-	-	-	-	-	-	896	-	896
Website	-	-	-	-	-	-	-	4,410	-	4,410
Office Supplies	-	-	-	-	-	-	-	3,758	-	3,758
Student & Member Awards	-	-	-	-	-	-	-	-	7,107	7,107
Fundraising Expense	-	-	-	-	-	-	-	-	1,602	1,602
Duplicating Expense	-	-	-	-	-	-	-	-	-	-
Chapter Expenses	-	-	-	-	-	-	-	-	1,774	1,774
Other Committee Expenses	-	-	-	-	-	-	-	-	1,000	1,000
Miscellaneous Office	-	-	-	-	-	-	-	110	-	110
Total	$ 219,005	$ 18,332	$ 12,925	$ 539	$ 22,396	$ 18,510	$ 570	$ 73,798	$ 11,483	$ 377,558

See Accompanying Notes.

LABOR AND EMPLOYMENT RELATIONS ASSOCIATION
STATEMENTS OF CASH FLOWS
FOR THE YEARS ENDED DECEMBER 31, 2021 AND 2020

	2021	2020
CASH FLOWS FROM OPERATING ACTIVITIES		
Change in Net Assets	$ 37,437	$ 56,878
Adjustments to Reconcile Change in Net Assets to Net Cash Provided by (Used in) Operating Activities:		
Depreciation Expense	2,079	1,639
Net Unrealized (Gains) Losses on Investments	8,749	22,303
(Increase) Decrease in Operating Assets:		
Accounts Receivable, Net	(1,406)	-
Prepaid Expenses	2,991	8,983
Inventory	313	579
Increase (Decrease) in Operating Liabilities:		
Accounts Payable	(750)	750
Accrued Liabilities	(2,394)	7,733
Dues Collected in Advance	2,931	2,197
Subscriptions Collected in Advance	(3,581)	(648)
Deferred Chapter Dues	(2,300)	1,900
Other Deferred Revenue	31,375	(14,000)
Net Cash Provided by (Used in) Operating Activities	75,444	88,314
CASH FLOWS FROM INVESTMENT ACTIVITIES		
Donated Investments	-	(41,980)
Reinvested Interest, Dividends, and Capital Gains	(3,383)	(3,911)
Purchases of Property and Equipment	(2,643)	-
Net Cash Provided by (Used in) Investment Activities	(6,026)	(45,891)
CASH FLOWS FROM FINANCING ACTIVITIES		
None	-	-
Net Cash Provided by (Used in) Financing Activities	-	-
Net Increase (Decrease) in Cash and Cash Equivalents	69,418	42,423
Beginning Cash and Cash Equivalents	325,553	283,130
Ending Cash and Cash Equivalents	$ 394,971	$ 325,553

See Accompanying Notes.

LABOR AND EMPLOYMENT RELATIONS ASSOCIATION
NOTES TO FINANCIAL STATEMENTS
DECEMBER 31, 2021 AND 2020

NOTE 1 – NATURE OF ACTIVITIES AND SIGNIFICANT ACCOUNTING POLICIES

Organization and Nature of Activities

The Labor and Employment Relations Association (the Organization) was founded in 1947 to encourage research in all aspects of the field of labor, employment, and the workplace. It is a nonprofit scholarly association of academic, labor, business, and neutral communities committed to the full discussion and exchange of ideas between and among its broad constituencies through meetings, publications, and its various electronic listservs and websites. The Organization's National Office is located in Champaign, Illinois and serves the association by planning conferences and meetings and publishing the various researches of its members. The main sources of support and revenues for the Organization are contributions, membership dues and subscriptions, and meeting income.

Subsequent Events

Subsequent events have been evaluated through September 22, 2022, which is the date the financial statements were available to be issued.

Accrual Basis of Reporting

The Organization has chosen to report on the accrual basis of accounting. Accordingly, revenue is recognized when earned and expenses are recognized when incurred in conformity with accounting principles generally accepted in the United States of America (GAAP). The financial statements are presented in accordance with Financial Accounting Standards Board (FASB) Accounting Standards Codification (ASC) 958 dated August 2016, and the provisions of the American Institute of Certified Public Accountants (AICPA) "Audit and Accounting Guide for Not-for-Profit Organizations" (the "Guide").

Estimates

The preparation of financial statements in conformity with GAAP requires management to make estimates and assumptions that affect certain reported amounts and disclosures. Accordingly, actual results could differ from those estimates.

Cash and Cash Equivalents

For purposes of reporting cash flows, cash and cash equivalents include all cash and highly liquid investments acquired with an original maturity date of three months or less. As of December 31, 2021 and 2020, $65,796 and $67,832 of cash and cash equivalents are designated by the board and are subject to board-imposed stipulations. As of December 31, 2021 and 2020, $28,876 and $23,031 of cash and cash equivalents are temporarily restricted and are subject to donor stipulations.

Investments

Investments consist of intermediate term bond funds, Bayer AG (BAYRY) stock, and money market funds. Net appreciation (depreciation) in the fair value of investments, which consists of the realized gains or losses and the unrealized appreciation (depreciation) on those investments, is presented in the statement of activities in accordance with donor restrictions as investment income. Investment income is presented net of investment fees. The average cost method is primarily used to determine the basis for computing realized gains or losses.

Accounts Receivable

Accounts receivable are recorded primarily for outstanding invoices for membership dues and subscriptions. An allowance for doubtful accounts is based on an analysis of expected collection rates determined from experience. The Organization had an allowance for doubtful accounts of $0 and $0 at December 31, 2021 and 2020, respectively.

Inventory

The Organization's inventory of directories, research volumes, proceedings, and perspective magazines is carried at the lower of cost and market value. Cost is determined on the basis of first in – first out.

Property and Equipment

Property and equipment expenditures in excess of $1,000 are capitalized at cost. Donated property and equipment are capitalized at estimated cost or fair market value at the time of donation. Depreciation of the assets is computed using the straight-line method over their estimated useful lives. The range of estimated useful lives by type of asset is as follows:

Furniture and Equipment	3 - 7 years

Net Assets

Net assets of the Organization and changes therein are classified and reported as follows:

Net Assets without Donor Restrictions - Net assets that are not subject to donor-imposed restrictions and may be expended for any purpose in performing the primary objectives of the Organization. The Organization's board may designate assets without restrictions for specific operational purposes from time to time.

Net Assets with Donor Restrictions - Net assets subject to stipulations imposed by donors, and grantors. Some donor restrictions are temporary in nature; those restrictions will be met by actions of the Organization or by the passage of time. Other donor restrictions are perpetual in nature, where by the donor has stipulated the funds be maintained in perpetuity. As of December 31, 2021 and 2020, the Organization had no donor restrictions that were perpetual in nature.

Contributions

Unconditional contributions are recognized when pledged and recorded as net assets without donor restrictions or net assets with donor restrictions, depending on the existence and/or nature of any donor-imposed restrictions. Conditional promises to give are recognized when the conditions on which they depend are substantially met. Gifts of cash and other assets are reported with donor restricted support if they are received with donor stipulations that limit the use of the donated assets.

When a restriction expires, that is, when a stipulated time restriction ends or a purpose restriction is accomplished, net assets with donor restrictions are reclassified to net assets without donor restrictions and reported in the statement of activities as net assets released from restrictions. Donor-restricted contributions whose restrictions are met in the same reporting period are reported as net assets without donor restriction support. Contributions restricted for the acquisition of land, buildings, and equipment are reported as net assets without donor restriction upon acquisition of the assets and the assets are placed in service.

Contributed Services and Goods

Contributed services are reported as contribution revenue and as assets or expenses only if the services create or enhance a non-financial asset (for example property and equipment) or:

- Would typically need to be purchased by the Organization if the services had not been provided by contribution.
- Require specialized skills.
- Are provided by individuals with those skills (such as accounting, financial, construction, educational, electrical, legal, medical, and other services provided by accountants, investments advisers, contractors, teachers, electricians, lawyers, doctors, and other professional and craftspeople).

For the years ended December 31, 2021 and 2020, the value of contributed services meeting the defined requirements for recognition in the financial statements as outlined above were not material and have not been recorded on the financial statements. The Organization does receive free office space from the University of Illinois. However, the value for the use of this office space has not been reported on the financial statements.

Income Tax Status

The Organization is a nonprofit association that is exempt from federal income tax under Section 501(c)(3) of the Internal Revenue Code. In addition, the Internal Revenue Service has determined that the Organization is not a private foundation as defined in Section 509(a)(1) and Section 170(b)(1)(A)(vi) of the Code.

The Organization has evaluated its exposure resulting from uncertain income tax position and determined the exposure is not material to the financial statements. In addition, the Organization is not aware of any tax position for which a significant change is reasonably possible within the next 12 months. Therefore, these financial statements do not include a liability for uncertain tax positions. Upon recognition of a liability for an uncertain tax position, the Organization would recognize interest expense and penalties in operating expenses.

The Organization files information tax returns in the U.S. federal jurisdiction and the state of Illinois. Its federal and Illinois information tax returns prior to calendar year 2018 are closed. The Organization does not have any tax returns currently under examination by either the Internal Revenue Service (IRS) or any U.S. state jurisdiction.

Membership Dues and Advance Subscriptions Collected

Membership dues and subscriptions are assessed and recognized as revenue based on the life of the dues or subscription.

Functional Allocation of Expenses

The costs of providing the various programs and other activities have been summarized on a functional basis in the statements of activities. The statements of functional expenses present the natural classification detail of expenses by function. Accordingly, certain costs have been allocated among the programs and supporting services benefited.

Expenses which are easily and directly associated with a particular program or supporting service are charged directly to that functional area. Compensation and payroll taxes & fringes have been allocated on the basis of estimates of time and effort.

NOTE 2 – ARRANGEMENTS WITH THE UNIVERSITY OF ILLINOIS

The Organization moved its offices to the University of Illinois at the end of 1999. Under an arrangement with the University, the employees of the Organization are employed by the University. The employees' pension and benefits are part of the University's plans. The Organization then reimburses the University monthly for the cost of its employees.

The University of Illinois holds some cash for the Organization. These "claim on cash" balances were $20,584 and $39,174 as of December 31, 2021 and 2020, respectively. These balances have been included with cash and cash equivalents.

NOTE 3 – CONCENTRATIONS OF CREDIT RISK

Financial instruments that potentially subject the Organization to credit risk consist principally of checking accounts, money markets accounts, and certificates of deposits at financial institutions. However, management continuously monitors the Organization's balances at financial institutions. The balances at each bank as of December 31, 2021 and 2020 were insured by the Federal Deposit Insurance Corporation (FDIC) up to $250,000. As of December 31, 2021, $84,767 of the bank balance was not insured by FDIC. As of December 31, 2020, the total bank balance was fully insured. The "claim on cash" held with the University of Illinois as stated in Note 2 is not insured since it is not held in a financial institution. In addition, the Organization had undeposited funds of $5,834 and $2,124 at December 31, 2021 and 2020, respectively. Since these funds were not yet deposited with a financial institution it was not covered by FDIC.

NOTE 4 – INVESTMENTS

FASB Codification 820, Fair Value Measurements, establishes a framework for measuring fair value. That framework provides a fair value hierarchy that prioritizes the inputs to valuation techniques used to measure fair value. The hierarch gives the highest priority to unadjusted quoted prices in active markets for identical assets or liabilities (Level 1 measurements) and the lowest priority to unobservable inputs (Level 3 measurements). The three levels of the fair value hierarchy under FASB Codification 820 are described below:

> Level 1 – Inputs to the valuation methodology are based on unadjusted quoted prices for identical assets or liabilities in active markets that the Organization has the ability to access.
>
> Level 2 – Inputs to the valuation methodology include quoted prices for similar assets or liabilities in active markets, quoted prices for identical or similar assets or liabilities in inactive markets, inputs other than quoted prices that are observable for the asset or liability, and inputs that are derived principally from or corroborated by observable market data by correlation or other means. If the asset or liability has a specified (contractual) term, the Level 2 input must be observable for substantially the full term of the asset or liability.
>
> Level 3 – Inputs to the valuation methodology are unobservable and significant to the fair value measurement.

The asset's and liability's fair value measurement level within the fair value hierarchy is based on the lowest level of any input that is significant to the fair value measurement. Valuation techniques used need to maximize the use of observable inputs and minimize the use of unobservable inputs.

The following is a description of the valuation methodologies used for assets measured at fair value. There have been no changes in the methodologies used at December 31, 2021 and 2020.

Intermediate Term Bond Funds:	Valued at the NAV of shares held by the Organization at year-end based on readily determinable fair values, which are published daily and are the basis for current transactions.
Bayer AG (BAYRY) Stock:	Valued at the stated share price at year-end.
Money Market Funds:	Valued at cost plus accrued interest.

The preceding methods described may produce a fair value calculation that may not be indicative of net realizable value or reflective of future fair values. Furthermore, although the Organization believes its valuation methods are appropriate and consistent with other market participants, the use of different methodologies or assumptions to determine the fair value of certain financial instruments could result in different fair value measurements at the reporting date.

The following table sets forth by level, within the fair value hierarchy, the Organization's assets at fair value as of December 31, 2021 and 2020:

Assets at Fair Value as of December 31, 2021

	Level 1	Level 2	Level 3	Total
Without Donor Restrictions				
Bayer AG (BAYRY)	$ 10,608	-	-	$ 10,608
Money Market Funds	756	-	-	756
Designated by the Board				
Intermediate Term Bond Funds	29,129	-	-	29,129
Bayer AG (BAYRY)	42,432	-	-	42,432
Money Market Funds	3,030	-	-	3,030
Total	$ 85,955	-	-	$ 85,955

Assets at Fair Value as of December 31, 2020

	Level 1	Level 2	Level 3	Total
Without Donor Restrictions				
Bayer AG (BAYRY)	$ 11,872	-	-	$ 11,872
Money Market Funds	425	-	-	425
Designated by the Board				
Intermediate Term Bond Funds	29,834	-	-	29,834
Bayer AG (BAYRY)	47,488	-	-	47,488
Money Market Funds	1,702	-	-	1,702
Total	$ 91,321	-	-	$ 91,321

Net investment earnings for the years ended December 31, 2021 and 2020 are summarized as follows:

	2021	2020
Dividends	$ 2,966	$ 3,705
Capital Gains Distributions	417	206
Net Unrealized Gains (Losses)	(8,749)	(22,303)
Total Investment Income	$ (5,366)	$(18,392)

This investment income, including unrealized gains and losses, are being reported on the statements of activity as investment income.

NOTE 5 – PROPERTY AND EQUIPMENT

As of December 31, 2021, Property and Equipment consists of:

Furniture and Equipment	$ 21,303
Less: Accumulated Depreciation	(18,011)
Property and Equipment, Net	$ 3,292

As of December 31, 2020, Property and Equipment consists of:

Furniture and Equipment	$ 18,660
Less: Accumulated Depreciation	(15,932)
Property and Equipment, Net	$ 2,728

Depreciation expense for the years ended December 31, 2021 and 2020 was $2,079 and $1,639, respectively.

NOTE 6 – NET ASSETS - BOARD DESIGNATED

Susan C. Eaton Scholar-Practitioner Memorial Fund

The Organization set up a memorial fund in honor of an author of a 1998 "Perspectives on Work" article who died on December 30, 2003. At its June 1, 2004 meeting, the Organization approved the establishment of an annual Susan C. Eaton Scholar-Practitioner award and grant to be paid to one or more qualified scholar researchers in even-numbered years or practitioners in odd-numbered years doing research in the labor and employment relations or related field. The Organization's Executive Board directed that 10% of any gifts received each year are undesignated and can be used for administrative expenses. As of December 31, 2021 and 2020, these designated funds totaled $20,779 and $20,779, respectively.

Kochan-Sleigh Best Dissertation Award Fund

On March 27, 2006, the Organization set up a designated fund with contributions from a member and matching contributions from General Electric. The fund is designated for a minimum of fifteen years and may be used to pay for the $1,000 annual best dissertation award and plaque. After the fifteen-year period, the fund may be continued or the amount remaining in the fund may be undesignated and become available to the Organization for unrestricted purposes. The Organization's Executive Board directed that 10% of any gifts received each year are undesignated and can be used for administrative expenses. As of December 31, 2021 and 2020, these designated funds totaled $36,439 and $37,439, respectively.

Jim Scoville Award Fund

On January 9, 2009, the Organization set up a designated fund with a $10,000 contribution from the University of Minnesota's Industrial Relations Center to honor a member and retiring professor. The fund was established to pay an annual award of $500 and a plaque for best paper on international and comparative employment issues. The Organization's Executive Board directed that 10% of any gifts received each year are undesignated and can be used for administrative expenses. As of December 31, 2021 and 2020, these designated funds totaled $2,796 and $3,296, respectively.

John T. Dunlop Public Policy Fund

On May 8, 2010, the Organization established a designated fund to subsidize a named John T. Dunlop Public Policy Session at future National Policy Forums and/or at the Organization annual meetings. The Organization's Executive Board directed that 10% of any gifts received each year are undesignated and can be used for administrative expenses. This fund will be so named for five or ten years, at which time the Organization's Executive Board will review the fund to determine whether to continue or modify the fund. As of December 31, 2021 and 2020, these designated funds totaled $18,326 and $19,005, respectively.

Gladys and Walter Gershenfeld Diversity & Inclusion Fund

On May 8, 2010, the Organization established a designated fund to honor the memory of long-time members and past presidents Walter and Gladys Gershenfeld for the purpose of supporting the Organization's electronic and print publications. The fund will be so named for ten years, at which time the Organization will review the fund's purpose and uses, and whether to continue or modify the fund. The Organization's Executive Board directed that 10% of any gifts received each year are undesignated and can be used for administrative expenses. As of December 31, 2021 and 2020, these designated funds totaled $16,585 and $17,147, respectively.

Quasi-Endowment Fund

During 2019, the Organization established a quasi-endowment fund for the purpose of supporting the Organization's general operations. The Organization's Executive Board directed that 20% of any gifts received each year are undesignated and can be used for administrative expenses. Funds designated by the Board of Directors to function as an endowment are voluntary and may be reversed by the Board of Directors at any time. Accordingly, they are reported as part of the net assets without donor restrictions. The Organization's Board designated endowment net assets exist as a permanent investment pool to enhance and sustain the operations of the Organization so that it can continue to be involved in the full discussion and exchange of ideas between and among its broad constituencies through meetings, publications, and its various electronic listservs and websites. As of December 31, 2021 and 2020, these designated funds totaled $45,462 and $49,190, respectively.

NOTE 7 – NET ASSETS WITH DONOR RESTRICTIONS

Net assets with donor restrictions at December 31, 2021 and 2020 are restricted for the following purposes or period:

	2021	2020
Subject to expenditures for specified purpose:		
Administer the Business of the Deans, Chairs, and Directors Committee (DCDC)	$ 8,320	$ 8,120
Administer the PhD Student Consortium	20,556	14,911
Total Net Assets with Donor Restrictions	$ 28,876	$ 23,031

NOTE 8 – ENDOWMENT

Board designated net assets at December 31, 2021 consist of an endowment fund established in 2019 for the purpose of supporting operations of the Organization. The Organization has adopted the provisions of FASB ASC 958-205-45, *Not-for-Profit Entities*, including *Net Asset Classification of Funds Subject to an Enacted Version of the Uniform Prudent Management of Institutional Funds Act, and Enhanced Disclosures for All Endowment Fund.* ASC 958 provides guidance on the net asset classification of donor restricted endowment funds for a non-for-profit organization that is subject to an enacted version of the State Prudent Management of Institutional Funds Act (SPMIFA) and also requires disclosures about endowment funds, both donor restricted and board designated endowment funds.

Interpretation of Relevant Law

The Organization has interpreted SPMIFA as requiring the preservation of the fair value of the original gift as of the gift date of the donor-restricted endowment funds absent explicit donor stipulations to the contrary. As a result of this interpretation, the Organization classifies as permanently restricted net assets (a) the original value of gifts donated to the permanent endowment, (b) the original value of subsequent gifts to the permanent endowment, and (c) accumulations to the permanent endowment made in accordance with the direction of the applicable donor gift instrument at the time the accumulation is added to the fund. The remaining portion of the donor-restricted endowment fund that is not classified as permanently restricted net assets is classified as net assets with donor restrictions - restricted for specified purpose until those amounts are appropriated for expenditures by the Organization in a manner consistent with the standard of prudence prescribed by SPMIFA.

Additionally, in accordance with SPMIFA, the Organization considers the following factors in making a determination to appropriate or accumulate donor-restricted endowment funds:

(1) The duration and preservation of the fund
(2) The purpose of the Organization and the donor-restricted endowment fund
(3) General economic conditions
(4) The possible effect of inflation or deflation
(5) The expected total return from income and the appreciation of investments

(6) Other resources of the Organization
(7) The investment policies of the Organization.

Endowment Net Asset Composition and Changes in Net Assets
The composition of endowment net assets and the changes in endowment net assets as of December 31, 2021 and 2020 are as follows:

	Without Donor Restrictions	
	2021	2020
Endowment Net Assets, Beginning of Year	$ 49,190	$ 32,448
Member Contributions	-	33,584
Investment Income	(3,728)	(16,842)
Endowment Net Assets, End of Year	$ 45,462	$ 49,190

Return Objectives and Risk Parameters
The primary objectives in the investment management for the portfolio assets is to provide ongoing stability and income for operations, preserving gift assets without subjecting them to substantial risk, and providing additional real growth through new gifts. The mix of investments in the portfolio is examined periodically.

Spending Policy
Spending from the fund should not exceed two to five percent annually depending on growth, general economic conditions, and organizational needs. The spending policy should be reviewed every five to ten years.

NOTE 9 – LIQUIDITY AND AVAILABILITY OF FINANCIAL ASSETS

The following reflects the Organization's financial assets as of the statements of financial position date, reduced by amounts not available for general use because of contractual or donor-imposed restrictions within one year of the statement of financial position date. Amounts available include donor restricted amounts that are available for general expenditure in the following year.

	2021	2020
Fiscal Assets at Year-End		
Cash and Cash Equivalents	$ 394,971	$ 325,553
Investments	85,955	91,321
Accounts Receivable, Net	1,406	-
Less Contractual or Donor-Imposed Restrictions		
Board Designated Funds	(140,387)	(146,856)
Donor Restrictions for Specific Purposes	(28,876)	(23,031)
Financial Assets Available to Meet Cash Needs for General Expenditure Within One Year	$ 313,069	$ 246,987

As part of the Organization's liquidity management, it has a policy to structure its financial assets to be available as its general expenditures, liabilities, and other obligations come due. Excess cash is generally held in checking accounts, money markets accounts, and certificates of deposits until it is required for operational use.

NOTE 10 – COMMITMENTS

The Organization signed a contract with the MotorCity Casino Hotel in Detroit on November 7, 2019 to hold the 73rd Annual Meeting from June 1st through the 7th, 2021. Due to COVID-19, the Annual Meeting went virtual in 2020 and stayed virtual for 2021. As a result, on September 10, 2020, the Organization did an addendum to the contract with the MotorCity Casino Hotel in Detroit to change the Annual Meeting date to May 31st through June 6th, 2022. All other conditions of the contract including concessions and meeting space apply as originally negotiated. If this contract was cancelled at December 31, 2021 and 2020, the Organization would have owed $40,437 and $40,437, respectively, to the MotorCity Casino Hotel in Detroit.

NOTE 11 – PENSION PLAN AND RETIREMENT CONTRIBUTIONS

The University of Illinois (the University) contributes to the State Universities Retirement System of Illinois (SURS), a cost-sharing multiple-employer defined benefit plan with a special funding situation whereby the State of Illinois (the State) makes substantially all actuarially determined required contributions on behalf of the participating employers. SURS was established July 21, 1941 to provide retirement annuities and other benefits for staff members and employees of state universities, certain affiliated organizations, and certain other state educational and scientific agencies and for survivors, dependents, and other beneficiaries of such employees. SURS is considered a component unit of the State of Illinois' financial reporting entity and is included in the State's financial reports as a pension trust fund. SURS is governed by Section 5/15, Chapter 40 of the *Illinois Compiled Statutes*. SURS issues a publicly available financial report that includes financial statements and required supplementary information. That report may be obtained by accessing the website at www.SURS.org.

Benefits Provided. A traditional benefit plan was established in 1941. Public Act 90-0448 enacted effective January 1, 1998, established an alternative defined benefit program known as the portable benefit package. The traditional and portable plan Tier 1 refers to members that began participation prior to January 1, 2011. Public Act 96-0889 revised the traditional and portable benefit plans for members who begin participation on or after January 1, 2011, and who do not have other eligible Illinois reciprocal system services. The revised plan is referred to as Tier 2. New employees are allowed six months after their date of hire to make an irrevocable election. A summary of the benefit provisions can be found in the SURS' comprehensive annual financial report (CAFR) notes to the financial statements.

Contributions. The State of Illinois is primarily responsible for funding SURS on behalf of the individual employers at an actuarially determined amount. Public Act 88-0593 provides a statutory funding plan consisting of two parts: (i) a ramp-up period from 1996 to 2010 and (ii) a period of contributions equal to a level percentage of the payroll of active members of SURS to reach 90 percent of the total actuarial accrued liability by the end of fiscal year 2045. Employer contributions from "trust, federal, and other funds" are provided under Section 15-155(b) of the Illinois Pension Code and require employers to pay contributions which are sufficient to cover the accruing normal costs on behalf of applicable employees. The employer normal cost was 13.02 percent of employee payroll during the period of July 1, 2019 to June 30, 2020. The employer normal cost was 12.70 percent of employee payroll during the period of July 1, 2020 to June 30, 2021. The employer normal cost was 12.32 percent of employee payroll during the period of July 1, 2021 to June 30, 2022. The normal cost is equal to the value of current year's pension benefit and does not include any allocation for the past unfunded liability or interest on the unfunded liability. Plan members are required to contribute 8.0 percent of their annual covered salary. The contribution requirements of plan members and employers are established and may be amended by the Illinois General Assembly. During the years ended December 31, 2021 and 2020, the Organization reimbursed the University of Illinois $20,280 and $21,561, respectively, for the employer's payments to SURS for eligible employees.

In addition to providing pension benefits, the State of Illinois provides certain health, dental and life insurance benefits to annuitants. This includes annuitants of the Organization. Substantially all State employees, including the Organization's employees, may become eligible for postemployment benefits if they eventually become annuitants. Health and dental benefits include basic benefits for annuitants under the State's self-insurance plan and insurance contracts currently in force. Life insurance benefits for annuitants under age 60 are equal to their annual salary at the time of retirement; life insurance benefits for annuitants age 60 or older are limited to $5,000 per annuitant. Currently, the State does not segregate payments made to annuitants from those made to current employees for health, dental and life insurance benefits. These costs are funded by the State and are not an obligation of the Organization.

Employees of the Organization may also elect to participate in several tax deferred annuity plans and defined contribution plans. These are single employer plans under which benefits are provided to participating employees through contracts issued to each individual. Participation and the level of employee contributions are voluntary. The Organization is not required to make contributions.

LERA gratefully acknowledges the continuing support of its 2021–22 Major Sponsors and Organizational Members

Major Sustaining Sponsors

International Union, UAW
UAW–Ford Joint Trust
Kaiser Permanente and the Alliance of Health Care Unions

Major University Sponsors

Cornell University, School of Industrial and Labor Relations
Florida International University, College of Business
Massachusetts Institute of Technology, Sloan School of Management/IWER
Michigan State University, School of Human Resources and Labor Relations
Pennsylvania State University, School of Labor and Employment Relations
Rutgers University, School of Management and Labor Relations
University of Illinois Urbana-Champaign, School of Labor & Employment Relations

Organizational Members

American Federation of Teachers, AFL-CIO
American Train Dispatchers Association
Boston University, Questrom School of Business
Centre for Employment and Labour Relations Law / Melbourne Law School
Cornell University, Scheinman Institute on Conflict Resolution
Cornell University, School of Industrial and Labor Relations
Federal Mediation and Conciliation Service
Florida International University, College of Business
Guercio & Guercio, LLP
International Union, UAW
Kaiser Permanente
Massachusetts Institute of Technology, Sloan School of Management
Michigan State University, School of Human Resources and Labor Relations
Pennsylvania State University, School of Labor and Employment Relations
Portland General Electric
Portland State University
Rutgers School of Management and Labor Relations
Taiwan Institute of Labor, Occupational Safety and Health
UAW–Ford Joint Trust
U.S. Department of Agriculture
United Auto Workers International
United Auto Workers Region 9A
United Steelworkers of America
University of California, Los Angeles, Anderson School of Management
University of California, Los Angeles, Institute for Research on Labor and Employment
University of California-Berkeley, Center for Labor Research and Education
University of Illinois Urbana-Champaign, School of Labor and Employment Relations
University of Minnesota, Center for Human Resources and Labor Studies
University of Toronto, Centre for Industrial Relations and Human Resources
University of Wisconsin School for Workers
W. E. Upjohn Institute for Employment Research

For more information about Organizational Membership with the LERA, please visit the LERA website: https://lera.memberclicks.net/organizational-membership.

ALPHABETICAL LIST OF AUTHORS